WITHDRAWN
FROM NORTH YORK PUBLIC LIBRARY

THOMAS COOK
Travellers

PRAGUE

BY
LOUIS JAMES

Produced by AA Publishing

Written by Louis James

Original photography by Jon Wyand

Edited, designed and produced by AA Publishing. Maps © The Automobile Association 1994

Distributed in the United Kingdom by AA Publishing, Fanum House, Basingstoke, Hampshire, RG21 2EA.

The contents of this publication are believed correct at the time of printing. Nevertheless, the publishers cannot be held responsible for any errors or omissions or for changes in the details given in this guide or for the consequences of any reliance on the information provided by the same. Assessments of attractions, hotels, restaurants and so forth are based upon the author's own experience and, therefore, descriptions given in this guide necessarily contain an element of subjective opinion which may not reflect the publishers' opinion or dictate a reader's own experiences on another occasion. **We have tried to ensure accuracy in this guide, but things do change and we would be grateful if readers would advise us of any inaccuracies they may encounter.**

© The Automobile Association 1994

All rights reserved. No part of this publication may be reproduced, stored in a retrieval system, or transmitted in any form or by any means – electronic, photocopying, recording, or otherwise – unless the written permission of the publishers has been obtained beforehand. This book may not be lent, resold, hired out or otherwise disposed of by way of trade in any form of binding or cover other than that in which it is published, without the prior consent of the publisher.

A CIP catalogue record for this book is available from the British Library.

ISBN 0 7495 0696 2

Published by AA Publishing (a trading name of Automobile Association Developments Limited, whose registered office is Fanum House, Basingstoke, Hampshire RG21 2EA. Registered number 1878835) and the Thomas Cook Group Ltd.

Colour separation: BTB Colour Reproduction, Whitchurch, Hampshire

Printed by Edicoes ASA, Oporto, Portugal

Cover picture: *Týn Church, Old Town Square*
Title page: *Old Town Hall*
Above: *Stork House, Old Town Square*

Contents

About this Book

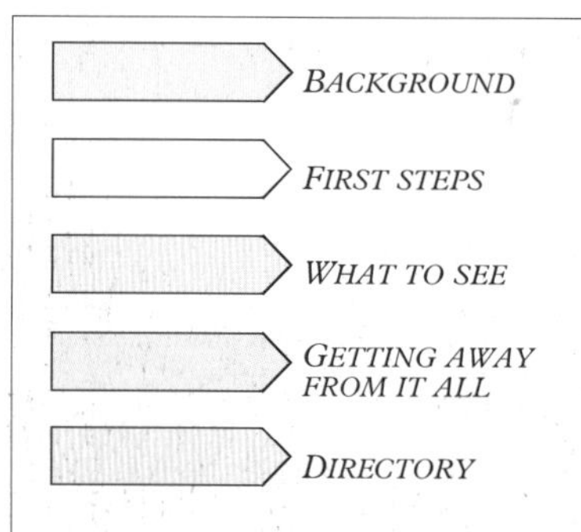

This book is divided into five sections, identified by the above colour coding.

CHANGES

Visitors to Prague should be aware that the city is still in a state of flux following the political changes of the last few years. Many museums and other sights have been, or are being, closed for restoration. This may mean some places are inaccessible to the public. A number of restaurants, coffee houses and the like may be in the process of changing hands, which means they are likely to close for refurbishment.

Telephone numbers

All 6- and 7-digit telephone numbers in Prague are changing to 8-digit numbers. At the time of printing, every effort has been made to change as many numbers as possible. If you have difficulty in dialling a number please ring 120 for Prague numbers and 121 for numbers in the Czech Republic.

Background gives an introduction to the city – its history, geography, politics, culture.

First Steps offers practical advice on arriving and getting around.

What to See is an alphabetical listing of places to visit, interspersed with walks and tours.

Getting Away From it All highlights places off the beaten track where it's possible to relax and enjoy peace and quiet.

Finally, the ***Directory*** provides practical information – from shopping and entertainment to children and sport, including a section on business matters.

Special highly illustrated features on specific aspects of the city appear throughout the book.

Prague, from the tower of the Old Town Hall

BACKGROUND

'Those who lie on the rails of history must expect to have their legs chopped off.'

RUDÉ PRAVÓ
Official organ of
the Communist Party, 1979

Introduction

The 'Golden City', the 'City of a Hundred Spires', the 'Rome North of the Alps': Prague (Praha) has inspired any number of flattering and romantic descriptions. It wears such laurels lightly, since its mixture of beauty and an atmosphere soaked in history speaks for itself.

Prague, it is often said, lies at the heart of Europe. Under its first ruling dynasty, the Přemyslids, and still more under the first three Luxembourg kings of Bohemia (1310–1419), it rose to be the focus of a great Central European empire. It reached its apogee of wealth and splendour under Charles IV (1346–78), who was also elected Holy Roman Emperor.

Strolling in Old Town Square

All this began to fall apart when the violent religious struggles of the 15th century between the followers of the reformer Jan Hus and Catholic rulers split the country. The Hussite wars also sharpened an age-old conflict between Germans (occupying many of the most influential offices of church and state) and native Slavs. In fact Prague had been a twin-cultured city, Slavic and Germanic, ever since Otakar II had invited Germans to colonise his new town of Malá Strana in 1257. It remained so up to the 20th century, German domination being the basis of the long period of Habsburg rule between 1526 and 1918. A third, deeply influential cultural element was supplied by the Jews, who formed 25 per cent of the city's population by 1700.

Despite being washed over by so many conflicts – dynastic, religious, racial – Prague has astonishingly remained one of the world's best preserved cities. Here, as nowhere else, you may drink your beer in one of dozens of Romanesque cellars, wander around some of the finest Gothic and baroque churches in Europe or track down a variety of exotic examples of art nouveau or cubist architecture. The people, too, are survivors, having found ways of living through wars and

LOCATOR MAP

oppression. This has given them a reputation for slyness on the one hand, resilience and courage on the other. The hero of a world-famous novel, *The Good Soldier Švejk*, represents the former quality; playwright and statesman Václav Havel, with his determination to 'live in truth', embodies the latter.

Prague today has come alive after 40 years of Communist gloom, even if the initial euphoria has been rapidly dissipated by bleak economic and political realities. Its marvellous musical

Mosaic in the recently restored Emmaus Monastery (Emauzy Klášter)

tradition has been given new life, artistic and cultural activity is breaking out in new directions and the restorers are at work on hundreds of beautiful but crumbling buildings and monuments. The city is on the move again; above all the young are racing to make up for lost time, hungry for new ideas, determined to succeed.

THOMAS COOK'S PRAGUE

When the state of Czechoslovakia was created after World War I, Thomas Cook began actively to promote tours to Prague and the surrounding countryside. In 1922 the Thomas Cook Travellers Gazette, *founded in 1851, carried an article promoting the spas and health resorts of Czechoslovakia and describing Prague as one of the most attractive cities on the Continent.*

History

8th century AD
Libuše, a legendary princess, has a vision of a city on the river, to be built where a man is found constructing the threshold (*práh*) to his house. In a subsequent vision she sees a young ploughman (*přemysl*) who will marry her and found the Přemyslid line.

Late 9th century
Duke Bořivoj I, the first historically verified Přemyslid, founds a citadel (Hradčany) on the Vltava.

Traditionally 929, but probably 935
Duke Wenceslas (later to become patron saint of Bohemia) is murdered by his brother, Boleslav I.

1085
Vratislav II is made first King of Bohemia by Emperor Henry IV.

1257
Otakar II settles German merchants and artisans in the new town of Malá Strana.

1306
The Přemyslid dynasty dies out.

1346
Prague's golden age begins with the accession of Charles IV of the Luxembourg dynasty. In 1355 he is crowned Holy Roman Emperor. Charles Bridge, St Vitus's Cathedral, Karlštejn and the Týn Church are built.

1348
Nové Město and Charles University (Karolinum) are founded.

1415
Jan Hus, the religious reformer, is burned for heresy at the Council of Constance.

1419
First defenestration of Prague: Hussites throw councillors from the windows of the New Town Hall.

1458–71
The Hussite king George of Poděbrady (Podiebrad) rules in Bohemia.

1483
Second defenestration of Prague: the mayor is thrown from the windows of the Old Town Hall.

A window on history: scene of the famous 1618 defenestration of Prague

1526
Ferdinand I becomes the first Habsburg king of Bohemia. He brings the Jesuits to Prague in 1556.

1576–1611
Rudolf II invites scholars, artists and astronomers to his Prague court, among them Tycho Brahe and Johannes Kepler.

1618
The third and most famous defenestration of Prague takes place: Protestant nobles throw Ferdinand II's councillors from the windows of Prague castle. This triggers the Thirty Years' War.

1620
The Bohemian Protestants are defeated at the Battle of the White Mountain, and the Counter-Reformation is driven forward. Fine churches and palaces are built but Prague is reduced to a backwater of the Habsburg Empire.

1740–80
Maria Theresa introduces enlightened reforms, but tries to expel the Jewish population from Prague. The city is twice attacked by Prussian King Frederick the Great.

1784
The four Prague towns of Hradčany, Malá Strana, Staré Město, and Nové Město are amalgamated. Joseph II allows freedom of religion and abolishes serfdom, but enforces German as the language of state.

1790–1848
The rise of Czech national consciousness culminates in the unsuccessful revolution of 1848.

1861
The first Czech Mayor of Prague is elected, leading to increasing tension between Czechs and Germans.

1918
At the end of World War I the Republic of Czechoslovakia is founded. The architect of Czech independence, Tomáš Masaryk, is its first president.

St Augustine, on the Charles Bridge

1939
The Nazis set up the 'Protectorate of Bohemia and Moravia'. Prague's Jews are sent to concentration camps.

1948
The Communist Party takes power. Non-Communist Jan Masaryk 'falls' from the windows of his Foreign Ministry (the fourth defenestration of Prague).

1968
The 'Prague Spring', led by Alexander Dubček, is crushed by the Warsaw Pact invasion.

1989
The Communist regime collapses on 10 December. On 29 December Václav Havel is elected President.

1990
Following elections in June a right-of-centre government is formed and begins the transformation of Czechoslovakia into a market economy.

1993
On 1 January the Czech and Slovak Republics formally separate. Prague becomes the capital of the new Czech Republic.

Politics

*I*n the 19th century Prague became the focus of the Czech national revival led by the distinguished historian František Palacký (1798–1876). Palacký refused to take part in the German National Assembly held in Frankfurt during the 1848 revolution against reactionary Habsburg government; instead he presided over a Pan-Slav Congress in Prague, an unmistakable signal that the days of German political and cultural domination were numbered.

The Republic

The man who realised Palacký's dream was Tomáš Garrigue Masaryk (1850–1937, see page 85). During World War I, Masaryk, in exile in the USA, persuaded the allies to recognise a new Czechoslovak state in the event of victory. On 28 October, 1918, the Czechoslovak Republic was declared in Prague with Masaryk as president. The Slovaks soon grew restive under what they regarded as Czech hegemony, and the German-speaking border area of Sudetenland had to be occupied after attempting to break away. These developments had ominous implications for the future integrity of Czechoslovakia.

Memorial to a modern-day hero: plaque commemorating student Jan Palach

World War II and Communism

Hitler occupied the Sudetenland in 1938 and the rest of Bohemia and Moravia the following year. The latter remained directly under Nazi control during the war, but the Slovaks were allowed an 'autonomous' state.

At the end of the war Edvard Beneš, the pre-war president, returned to his post and presided over the expulsion of most of the Sudeten Germans (nearly 2.5 million fled or were forced to leave). The Communist Party quickly infiltrated the organs of government, but it also enjoyed great popular support, not least because the 1938 betrayal of Czechoslovakia by Britain and France at Munich was still fresh in people's minds. In the 'Victorious February' of 1948 the Party consolidated its grip without taking up Stalin's offer of military assistance.

The Prague Spring

After 15 years, pressure for reform began to grow and the hardliners lost ground. In 1968 the Slovak Alexander Dubček emerged as leader, pledged to bring in 'Socialism with a human face'. The Russians unleashed the Warsaw Pact invasion of August 1968 and Gustáv Husák reimposed orthodoxy, surviving (latterly as president) until the Velvet Revolution of 1989 (see pages 12–13). Resistance to the invasion was non-violent and symbolic, most dramatically in the case of the student Jan Palach who committed suicide on Wenceslas Square by setting fire to himself. Some 150,000 people fled the country.

1989 and its consequences

A worsening economy, the more liberal regime of Mikhail Gorbachev in Russia and the opposition of Charter 77 (formed to monitor human rights) put increasing pressure on the regime, culminating in the 'Velvet Revolution' of November–December 1989 which swept Václav Havel to the presidency. Neither Havel nor his Prime Minister, Václav Klaus, were able to halt the subsequent break-up of Czechoslovakia into two separate states on 1 January 1993.

The new Czech Republic (which consists of Bohemia and Moravia) is a pluralist democracy with a government elected by proportional representation. Václav Havel was elected president (by parliament) for a five-year term, although with powers considerably reduced from those he believed he required.

The country is being subjected to a crash privatisation programme, and has enormous potential in areas such as tourism (chiefly to Prague). Its problems are inflation, a weak currency, a decimated industrial base, and the new phenomenon of unemployment (but not in Prague). A few people are doing very well, but many are disoriented and relatively impoverished.

Sign proclaiming the new Czech Republic

Prague today has two faces, one exuding self-confidence, one lined with worry. To many, freedom without a measure of prosperity to go with it seems like a sports car lacking an engine: nice to look at, but going nowhere. On the other hand, there is substantial inward investment; if the rest of Europe pulls out of recession, the prospects for the Czech Republic look promising.

THE VELVET REVOLUTION

After the crushing of the 'Prague Spring' in 1968, Czechoslovakia sank again under the oppression of Stalinism. Opposition began to surface with the formation of Charter 77. This monitored abuses of human rights outlawed by the Helsinki Agreements, which were signed by the Communist governments, with no intention of implementing them.

By the mid-1980s the Czechoslovak Stalinists found themselves increasingly undermined by Mikhail Gorbachev's policy of *perestroika*. Demands for more freedom came from young people, Charter 77 and the Catholic Church.

At the end of 1989 the organisations Civic Forum (in the Czech Republic) and People Against Violence (in Slovakia), were formed in response to police brutality against demonstrators on 17 November.

The following week vast crowds poured into Wenceslas Square every evening demanding the resignation of the government. On Friday evening Dubček addressed the crowds with Václav Havel, the leading personality of Civic Forum.

The Party tried forming a new government with puppet figures, but the tide was running against them. On 10 December the first cabinet since 1948 with Communists in a minority was announced, together with a promise of multi-party elections in June 1990. Posters immediately appeared with the slogan HAVEL NA HRAD! (Havel to the castle!) and demonstrations continued until he was unanimously elected president by the Federal Assembly on 29 December.

The 'Velvet Revolution', as it was dubbed, was not pushing at an open door: through massive peaceful demonstrations, coupled with the threat of a general strike, it defeated desperate and ruthless men. However, the deciding factor was the refusal of Gorbachev to save the puppet regime maintained by his predecessors. As soon as it was clear that the regime could not summon foreign tanks to save itself, it collapsed like a pack of cards.

Velvet diplomacy: Václav Havel, with Dubček (left)

Culture

Paradox and puzzle
The paradoxical nature of Prague and its people has constantly been reflected in the city's art, architecture and literature. Religious conflict has been linked to this from earliest times and has shaped the face of the city. Hussite mobs wrecked the interiors of churches, but the Jesuit-led Counter-Reformation filled Prague with sensual, drama-filled architecture and images. A pattern of victories and defeats has produced a many-layered culture, part of it always submerged. Prague has consequently been described as a 'metaphysical madhouse' and a place where 'the absurd is the paradoxical condition of human existence'.

St Nicholas Church in the Lesser Quarter

Art and architecture
Prague's rich heritage of architecture begins with the Romanesque buildings below street level and three rotunda churches from the 11th and 12th centuries. Gothic architecture dominated from the mid-13th century (Cathedral of St Vitus, Týn Church and St Agnes Convent). Renaissance style, imported from Italy under the first Habsburgs, is represented by several town houses and palaces, but above all by the Belvedere (1563) built by Ferdinand I.

With the Counter-Reformation baroque style conquered all, huge palaces and magnificent churches appearing in the 17th and 18th centuries (St Nicholas in Malá Strana, St James in Staré Město). This creative energy was manifest in painting in the works of such artists as Karel Škréta and Cosmas Asam, and sculpture reached a high point of expressiveness with Matthias Braun, Ferdinand Brokoff and others.

After a period of historicism, when architecture replicated earlier styles, Prague was swept by the sensuous lines of art nouveau at the end of the 19th century (see pages 28–9), followed by pioneering experiments with cubist architecture, unique to the city (see pages 40–1).

In the 20th century, Czech artists and sculptors began to look to the west. Alfons Mucha, the best known art nouveau painter, and the abstract artist František Kupka lived in Paris, while the sculptor František Bílek studied there. Between the wars the Devětsil (Nine Forces) movement produced Prague's own version of avant-garde art, architecture and literature.

Music

'Whoever is Czech is a musician' runs a proverb, and this was certainly the impression given in the baroque and romantic eras. Jan Stamic was the court composer in Mannheim, Antonín Rejcha in Paris was the teacher of Berlioz, Gounod, Franck and Liszt, and the Italians celebrated Josef Mysliveček as '*il divino boemo*' (the divine Bohemian), while an appreciative audience in Prague gave Mozart his greatest success with the première of his opera *Don Giovanni* (see pages 92–3).

The 19th century produced the outpouring of patriotic romantic music which is now part of every orchestral repertoire – such works as *Má Vlast* (My Homeland) by Bedřich Smetana (1824–84), the *Slavonic Dances* of Antonín Dvořák (1841–1904), or the *Glagolitic Mass* of Leoš Janáček (1854–1928).

Antonín Dvořák's piano – Vila Amerika

Literature and drama

Turn-of-the-century Prague produced a flowering of writing in German, most famously that of Franz Kafka (see pages 64–5). Czech writing was chiefly known abroad for the bleak realism of Karel Čapek and Jaroslav Hašek's comic masterpiece *The Good Soldier Švejk*; then came the contemporary wave of writers such as Václav Havel, Milan Kundera, Josef Škvorecký, Bohumil Hrabal and Ivan Klíma, with works of surreal humour, eroticism and bleak irony. The Nobel Prize-winning poet, Jaroslav Seifert, summed up the choice facing Czech writers in the last 40 years: 'When an ordinary person stays silent, it may be a tactical manoeuvre. When a writer stays silent, he is lying.'

Geography
(Topography and Climate)

PRAGUE

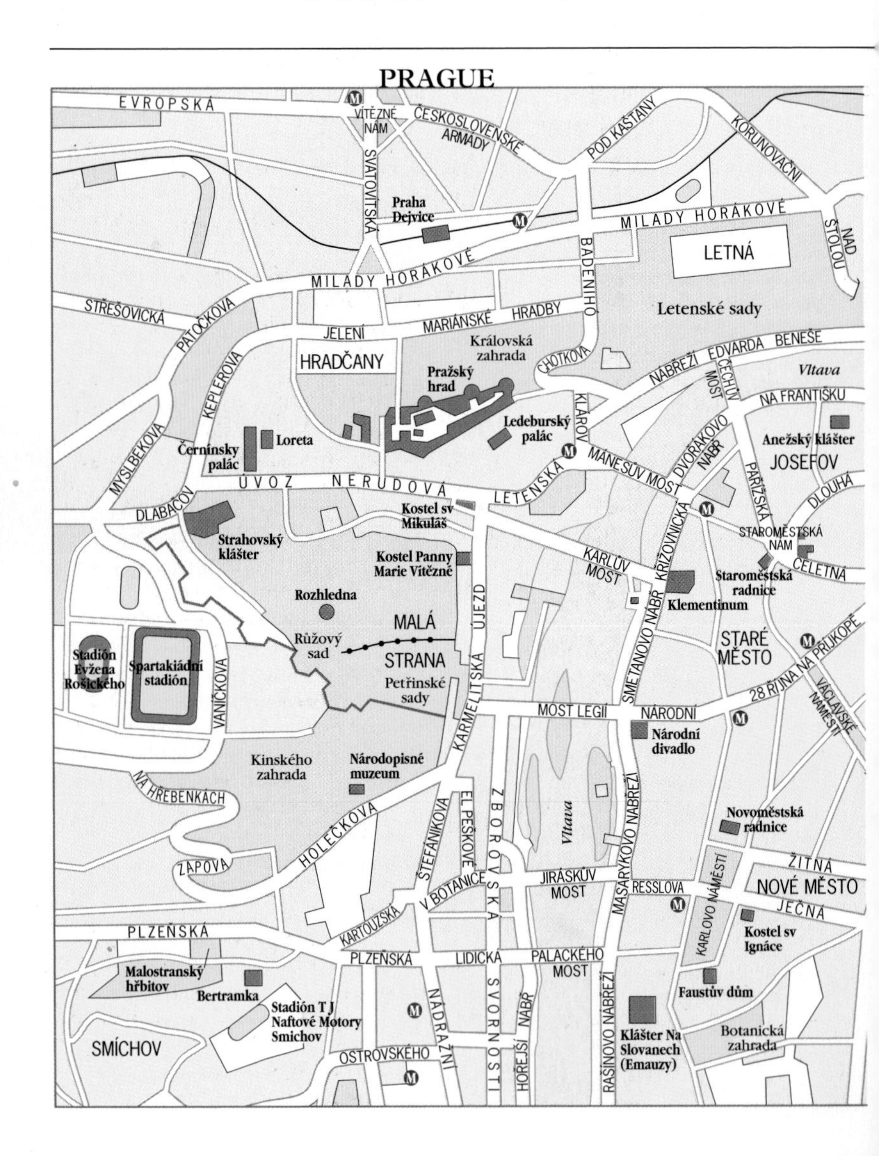

Prague lies 50 degrees 5 minutes north and 14 degrees 25 minutes east – a city in the centre of Europe, but part of Western European culture. Its topography is determined by the River Vltava. The 497sq km Prague conurbation stretches 48km along both banks of the river, a huge expansion from the modest town of 1883 (8.5sq km). Much of this increase is accounted for by the annexation of suburbs over the years. Historic Prague is a very small area.

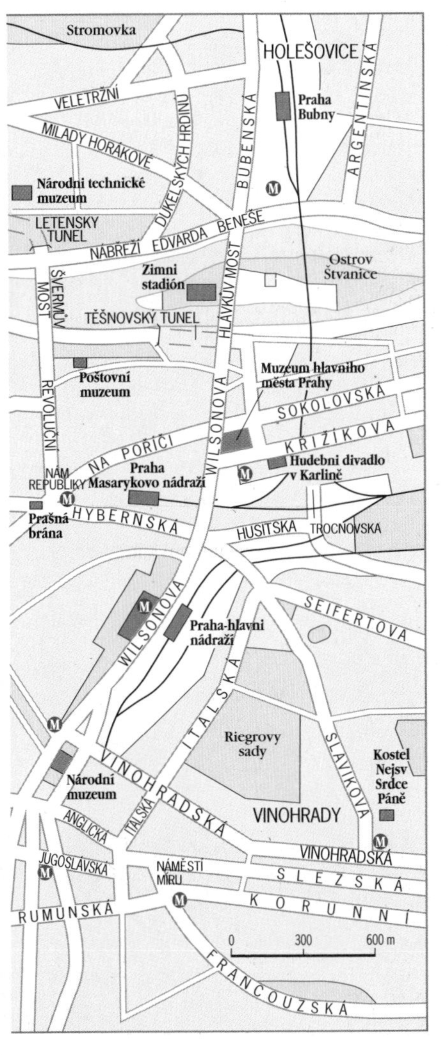

Strategic Prague

Like Rome, Prague is built on seven hills. On the left (western) bank of the Vltava the wedge-shaped plateau called Hradčany rises abruptly from a narrow ribbon of land along the shore. One of the three great cities of Central Europe (the other two being Vienna and Cracow), Prague lay on important trade routes crossing Europe from Germany, Poland, Russia and more exotic parts.

The toll levied on the Judith Bridge across the Vltava brought in revenue for the crown: the town benefited from the supply of goods and services. The building of the Charles Bridge was also a practical necessity, since flood waters had all but carried away its predecessor. While the original inhabitants of Prague had chosen the better protected slopes above the west bank for their dwellings, flooding on the east side continued until the late 13th century. The problem was partially solved by raising the street level in Staré Město (Old Town). Systematic regulation of the Vltava had to wait until the 19th century.

Climate

Prague's climate is determined by a mixture of oceanic and continental influences. The average temperature of 9°C implies a mild climate, but it can be very hot for long periods in summer, while in winter the thermometer stays at or below zero for three months.

Pollution in Prague

'Forests dead or dying, rivers dirtied, air unbreathable, soil choked with chemicals.' That is how one recent publication described the environmental disaster affecting the former Czechoslovakia. On top of acid rain, substandard water supplies and rivers incapable of supporting fish life, the country's air pollution, as measured by sulphur dioxide emissions per sq km, is the worst in Europe after that of East Germany. The problem in Prague itself is worst in winter as a result of 'inversion', when cold air is trapped in the Prague basin: the sulphur dioxide from factories and nitrogen oxide from automobiles and buses cannot escape. Dust, noise and even industrial stenches are other aspects of the 'golden city's' pollution. The burning of low-grade brown coal, an inefficient and toxic source of energy, is the greatest single pollutant. Allergies, chronic conditions of the respiratory tract and weakened immune systems are the result.

The new government attempts to do what can be done within the constraints of sparse budgets. Its declared priority is to replace brown coal as an energy source wherever possible. This means retaining a nuclear power plant which needs to be brought up to acceptable standards of safety. Neighbouring countries fear another Chernobyl, but most Czechs accept it as a *pis aller*.

Less controversial is the law requiring all cars sold on the Czech market to have a catalytic converter. In Prague and elsewhere recycling programmes have been started: separate rubbish containers have been introduced for white or coloured glass, and for paper. Regular monitoring of air pollution is now standard in the media. All this is a start, but the massive investment required for decommissioning or modernising polluting factories and cleaning up water supplies will take years to achieve.

Heavy traffic pollutes the city

First Steps

'Prague always had two faces. She was officially German and unofficially Czech. Or she was officially Czech, but unofficially she had within herself a German city, with its own schools, universities, cinemas, theatres, restaurants, coffee-houses, newspapers. She was officially Austrian, and unofficially anti-Austrian. She was officially Catholic and unofficially anti-Christian ....'

WILLY LORENZ,
To Bohemia with Love

First Steps

*T*he approaches to Prague, usually through acres of characterless concrete blocks or grey suburbs with belching smokestacks, hardly prepare you for the prize at journey's end. When finally you find yourself in the historic centre, it is as though a curtain has suddenly fallen away, revealing a city within a city, jewels of architecture held in a time capsule. The French writer André Breton aptly described this city as 'the magic metropolis of old Europe'.

First impressions

Despite its air of being suspended in time, Prague today is in a state of flux. After the Velvet Revolution of 1989 dramatic changes occurred almost overnight – some of them universally welcomed, some of them highly controversial. Uncontroversial has been the systematic restoration of old street names usurped by Communist mythology: the plaster saints of a cruel and treacherous regime have been gleefully swept into the dustbin of history. On the other hand the restitution law for property expropriated under Communism often leaves its

PRAGUE ENVIRONS

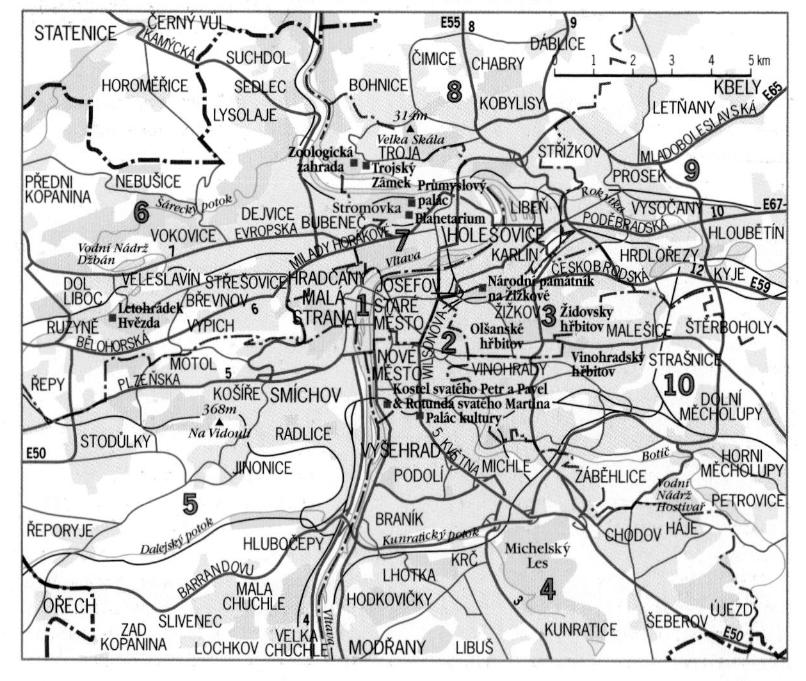

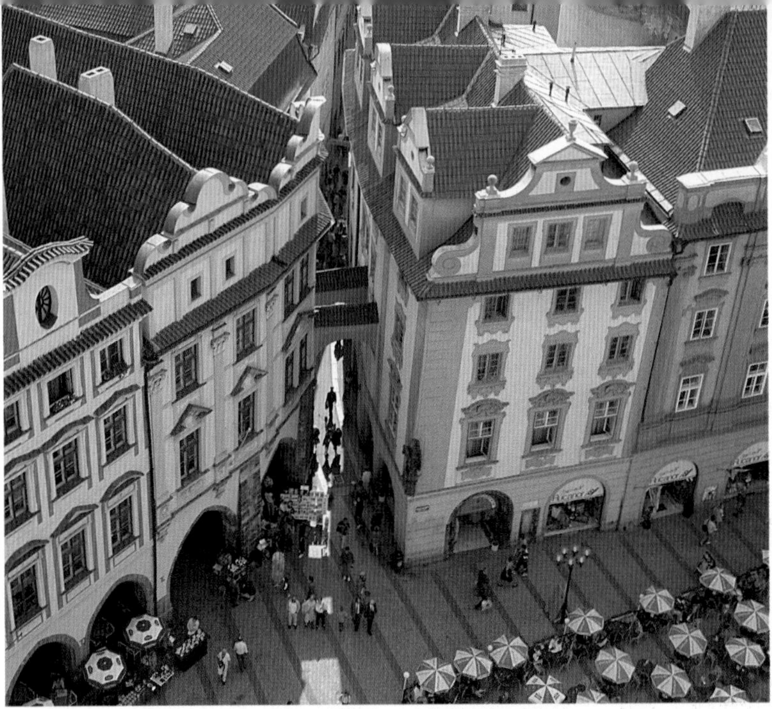

The narrow streets of the Old Town (Staré Město) are full of hidden delights

beneficiaries frustrated and its potential losers understandably resentful. Tenants fear eviction or soaring rents, while many owners discover they are unable to exploit or enjoy their reacquired properties.

Other shocks to the system will be even more evident to the visitor. Restaurants, shops and nightspots mushroom and disappear with bewildering rapidity. Luxury models of western cars glide around the city, competing for tarmac with elderly Škodas belching fumes. Security guards swagger in the entrances to expensive shops, while pensioners queue in corner groceries for increasingly expensive necessities.

The backdrop to such contrasting daily scenes is the cluster of diminutive, historic towns comprising old Prague. So rich in preserved architectural beauty are these areas, it is little wonder that most Praguers are irrepressible local patriots. Rows of baroque façades remind one of a film set (and are frequently used as such), while palaces and churches cram the narrow streets of Malá Strana (the Lesser Quarter) and of Hradčany. Here, and in Staré Město (the Old Town), many such streets are ancient cobbled alleyways from which cars have long been banned. In these backwaters Prague's uniquely ambivalent atmosphere of magic and menace can most powerfully be sensed. An exotic cast of emperors, warlords, religious fanatics, mystical rabbis and the peoples of three different races have left their mark on the ancient core of the city. The past is a babble of competing voices, the present a turbulent transition. Prague remains, as Egon Erwin Kisch described it, 'the market-place of sensations'.

When to go

Prague can be uncomfortably hot in high summer and very cold and raw in winter. The best time to visit the city (and Central Europe generally) is late spring or early autumn. It is not warm before April and gets rapidly colder after October.

Many Praguers traditionally leave town in summer and head for their dachas in the country. They now have an added incentive to do so, for the city is full of tourists in July and August. Although new facilities are being built and old ones expanded, it is still preferable to avoid these two months, as well as Whitsun and Easter. It should also be borne in mind that in winter a number of sights (especially palaces and their gardens) may be closed.

Getting around

From the point of view of the visitor both the topography of Prague and its public transport system are decidedly user-friendly. The main sights are in a compact area comprising the historic areas of the city (see **Areas of Prague**, page 24) and are ideally visited on foot. The heart of Staré Město is a pedestrian zone as is most of Wenceslas Square and the whole of the castle area.

Tram

The best way to see the city is by tram. They run along both banks of the Vltava and cross the bridges to traverse Malá Strana as far as the northern approach to the castle.

Metro

Three lines criss-cross the city and a fourth is planned. If you take a private room in the suburbs, check that it is close to the metro. Trains run every two minutes at rush hours, every six minutes off-peak. A single ticket entitles you to an hour's ride anywhere on the system. Remember that *výstup* means exit and *přestup* indicates an interchange with other lines and you should have few problems.

Prague's metro is clean, fast and cheap

Trams are an attractive alternative to the metro

Buses
Buses serve outlying areas and you are only likely to use them for excursions and a few specific city destinations.

Taxis
Taxis are plentiful, and cheap by western standards. Since the drivers know this, overcharging is common; it is always advisable to check that the meter is turned on (*zapněte taxametr, prosím* is the Czech for requesting this). If you are overcharged, ask for a receipt (*prosím dejte mi potrzení*) and write down the cabbie's number.

Car
Since the old city is compact and public transport runs up to the edge of its pedestrian zones, there is little point in driving around Prague. Moreover, as street parking is problematic and being towed away is the common fate of unwary tourists, it is best to leave your car in a secure garage. An epidemic of car theft in Central Europe since the collapse of the Iron Curtain shows no sign of abating.

For more information on public transport and driving see pages 182–3 and 188.

Manners and mores
Czechs attach great importance to the outward forms of courtesy: failure to greet your neighbours on the stairs, if you are living in a private apartment, could well be taken amiss. Similarly, if you share a café or restaurant table (and you may find you have to), a *dobrý den* (good day) when you sit down and a *na shledanou* (goodbye) when you leave are *de rigueur*.

Central European housewives are extremely house-proud; offer to remove your walking shoes on arrival at a flat. Society in this part of the world is still patriarchal, despite stirrings of emancipation. It will be some time before feminism is accepted and even longer before the commanding heights of business and politics are conquered by women.

The cost of living

Communist regimes offered job security and controlled prices to their shackled citizens. The arrival of democracy and free market policies has exacted a high toll in unemployment and rocketing prices. However, due to the continual devaluation of the Czech crown against hard currencies, Prague will seem cheap to most visitors: rooms in private houses, quite substantial meals and such things as concert tickets cost as little as half West European prices. For most Czechs the inflation rate and the new tax regime (VAT introduced in 1992) have meant a severe drop in living standards. The hardest hit are those on fixed incomes; but salaries remain abysmally low, while prices have continued to rise.

Areas of Prague

'Prague is actually composed of two parts. To the right of the Vltava is the Old Town – the Prague of Franz Kafka, the Prague of the Jewish community, the Prague of the Hussites, Czech Prague. To the left of the Vltava is the Lesser Quarter – the baroque, the Catholic Prague, the Prague of the Counter-Reformation with its palaces of the nobility, with its many churches and monasteries.'

Barbara Coudenhove Kalergi.

It was only in 1784 that Emperor Joseph II ordered the four historic towns of Prague – Hradčany and Malá Strana (the Lesser Quarter) on the west bank of the Vltava, Staré Město and Nové Město (the Old and New Towns) on the east bank – to be merged into a single municipality. Josefov, the Jewish ghetto, was amalgamated with Staré Město in 1850. Except for a large slice of Nové Město, these historic areas today constitute Prague's First District.

While most of historic Prague, and therefore most tourist sights are in Praha 1, the modern capital is administered in a total of 10 districts. These are the result of five further phases of expansion after 1784. The last growth of residential and industrial suburbs occurred in 1974.

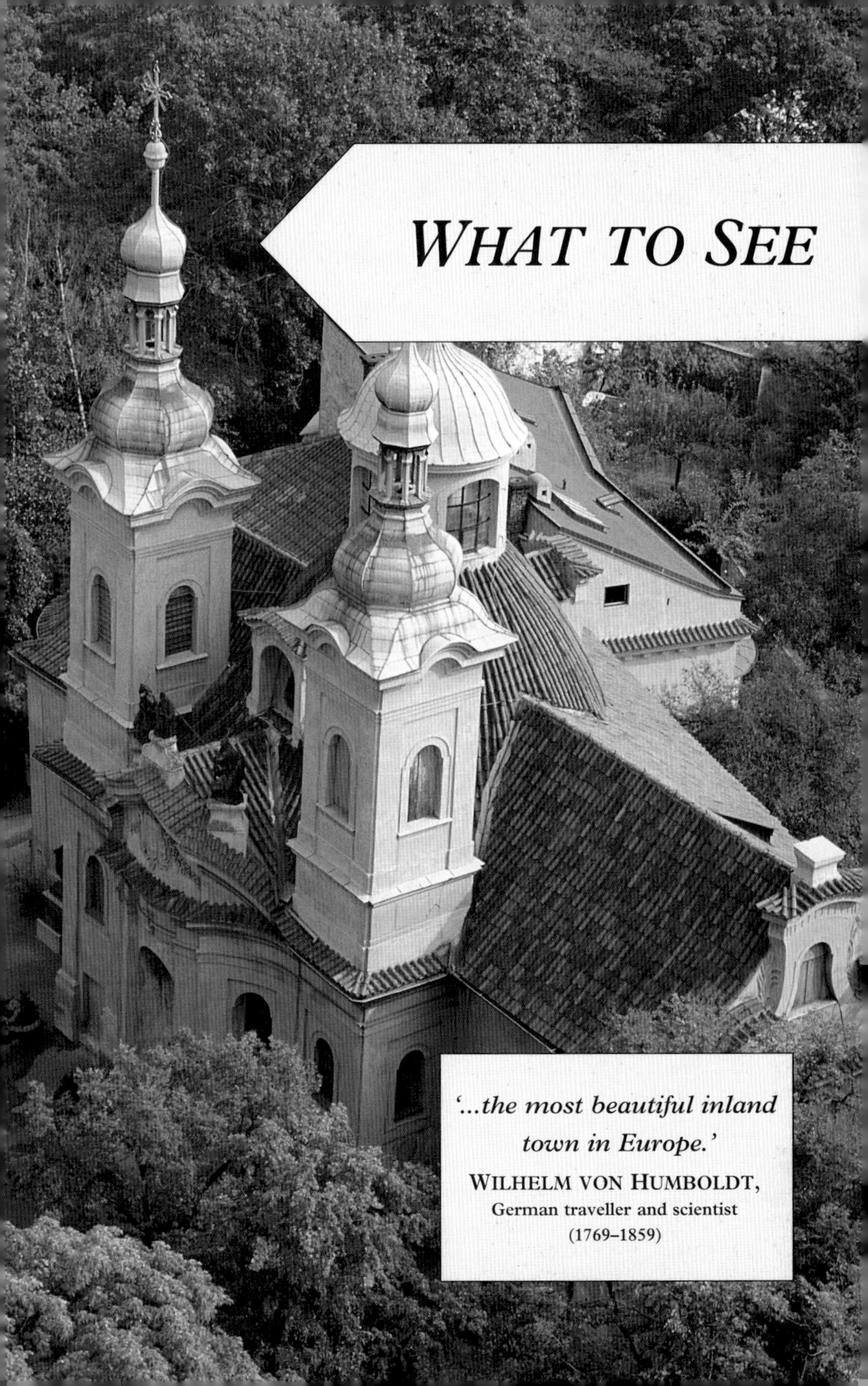

WHAT TO SEE

'...the most beautiful inland town in Europe.'

WILHELM VON HUMBOLDT,
German traveller and scientist
(1769–1859)

Anežský Klášter

(St Agnes' Convent)

Beginnings

The Convent of St Agnes is a particularly fine example of painstakingly restored early Gothic architecture. It was founded at the instigation of Anežka, the sister of Wenceslas I, in 1233; she preferred the relative freedom of life in a convent to the less attractive proposition of a dynastic marriage.

In 1235 Agnes (Anežka) became the first abbess of the new foundation, which was occupied by the mendicant order of Poor Clares. Some five years later the Franciscans, male counterparts of the Poor Clares, settled in a monastery next door. The whole complex was sometimes referred to as the 'Bohemian Assisi', after the Italian town where, only a few years earlier, the two orders were founded side by side. There was also a Minorite monastery near by, remains of which were only discovered in the course of 20th-century restoration.

History

Emperor Joseph II dissolved the convent in 1782, on the grounds that it served no useful purpose, and it was turned into an old people's home. In the succeeding hundred years it progressively decayed, becoming partly a slum and partly a rabbit warren of craftsmen's studios.

The courtyard of St Agnes' Convent

In the 1890s a patriotic fund was inaugurated to clean up the whole area and restore the buildings.

Both Agnes and Wenceslas are buried here, together with other members of the Přemyslid dynasty. Notwithstanding the convent's significance as a dynastic burial place (or perhaps because of it), it was vandalised during the Hussite wars (see page 37) and the inmates seem to have been compelled to leave. It was not until 1556 that it was reoccupied – by the Dominicans. However, the Poor Clares did return in 1627 and remained until 1782. Their first abbess had to wait until 1989 before she was canonised. That was on the eve of the Velvet Revolution and shortly before Pope John Paul II's historic visit to Prague.

The buildings

The cloister, dating to about 1260, is an open arcade around a square courtyard. It has a heavy vaulted ceiling characteristic of early Gothic. To the east of it, reached through a passage, is the most impressive of the surviving buildings, the Kostel sv Salvátora (Church of the Holy Saviour). An interesting feature of this French-influenced Gothic church is the relief portraits on the capitals of the arched entrance, apparently those of the kings and queens of the Přemyslid dynasty. The head over the salvation altar is thought by scholars to be that of St Agnes herself, watching over the nearby entrance to the royal crypt. The loveliest part of the church is the choir.

South of the Church of the Holy Saviour is the Kostel sv František (Church of St Francis), dating to about 1240 and a somewhat severe edifice built according to the puritanical architectural norms laid down by the Minorites. King Wenceslas I is buried in the church, which has only recently reacquired a roof and is now used for lectures and concerts.

The cloistered calm of Prague's medieval past

The convent also houses part of the Czech National Gallery's collections (see pages 90–1); 19th-century applied arts and painting are on the ground floor, and painting and sculpture from the same period on the first floor. Most of these works are imbued with the patriotic sentiment of the Czech national revival – not easy for everyone to appreciate. However, some of the landscapes and evocative scenes of turn-of-the-century Prague are particularly attractive.

U milosrdných 17. Staré Město, Praha I. Tel: 2481 0628. Open: Tuesday to Sunday 10am–6pm. Admission charge. Trams 5, 14 and 26 to Revolucní.

Art Nouveau Architecture

Art nouveau, a term first used in Paris in 1895, liberated architecture and the fine arts from rigid formalism and quotation. The natural world was its touchstone; it luxuriated in decorative flowing lines with floral patterns and exotic ornament.

When art nouveau arrived in Prague much of Central European architecture seemed frozen in time. Those who gave the big commissions – principally the state and the church – expected architects to adhere to the styles of the past. After a while this so-called 'historicism' began to degenerate into a sterile and pompous reproduction of motifs from pattern books. Art nouveau not only represented a new aesthetic approach, but also took advantage of advances in construction technology, using materials such as cast iron, steel and glass.

Josef Fanta's Central Station

Prague contains some of the most fascinating of art nouveau works, here usually labelled 'modern style' or *secesní* after the Viennese Secession movement. One of the most successful proponents of the new style in Paris was the Czech, Alfons Mucha (1860–1939), renowned for his enduringly popular posters. You can see some of his work in the Mayor's Hall, the highlight of a visit to Obecní Dům (the Community House on Náměstí Republiky, see pages 104–5), completed in 1911 and perhaps the most ambitious of Prague's Secessionist buildings.

EVROPA HOTEL (Europa Hotel)
This highly ornate hotel is essential viewing for anyone nostalgic about a world that died in 1914. It was built in 1906 by Alois Dryák and Bedřich Bendelmayer.
25 Václavské náměstí. Tel: 2422 8118 or 2422 8119. Café open: 7am–11pm. Metro to Muzeum or Můstek.

HANAVSKÝ PAVILÓN (Hanava Pavilion)
Built for the Paris World Exhibition of 1878, the pavilion is a historicist folly. It is the use of cast iron in the construction which gives it an art nouveau flavour.
Letenské sady. Tel: 32 57 92. Wine cellar open: daily noon–1am (disco). In summer – terrace café during daylight hours. Trams 2 and 17 to Čechův most then climb the steps up to the park.

Art nouveau window in St Vitus' Cathedral

HLAVNÍ NÁDRAŽÍ (Central Railway Station)

The art nouveau façade and concourse for the station were designed by Josef Fanta in 1909. This is one of the most monumental of Prague's art nouveau buildings

Wilsonova. Metro (line C) to Hlavní nádraží.

PETERKŮV DŮM (Peterka House)

This is a rather restrained example of art nouveau by Jan Kotěra, a pupil of the great Viennese Secessionist architect, Otto Wagner.

12 Václavské náměstí. Not open to the public. Metro to Muzeum or Můstek.

POJIŠTOVNA PRAHA (Prague Savings Bank) and TOPIČŮV DŮM (State Publishing House)

The buildings, nos 7 and 9 respectively, are neighbours on Národní. Note the lavish mosaic lettering above the windows, advertising the bank's various services, and the ceramic reliefs in the gable of the Topic publishing house. Both houses were designed by Oswald Polívka and built between 1907 and 1908 (see pages 122–3).

7 and 9 Národní. Not open to the public. Trams 6, 9, 18 and 22 to Národní divadlo.

PRŮMYSLOVÝ PALÁC (Industrial Palace)

Bedřich Münzberger's huge steel and glass construction, built for the Bohemia Jubilee Exhibition of 1891, is the first Prague building that recognisably embodies the spirit of art nouveau. It still has neo-baroque features, but the materials used were innovative.

Praha 7, Výstaviště. Open: May to September, 10am–5.30pm free, evenings – admission charge. Metro to Nádraží Holešovice. Trams 5, 12 and 17 to Výstaviště.

Those hungry for more can pick up at any bookshop an interesting booklet with colour illustrations entitled *Art Nouveau in Prague* by Petr Balajka.

The grandeur of Pařížska Street

THE VLTAVA

The Vltava (Moldau to the Germans) dominates Prague as few other rivers dominate a capital city. In the past whole sections of the population used to live from it, but it had less appealing roles as well: thieves and adulterers were suspended in its icy waters in wicker baskets from the Charles Bridge; John of Nepomuk, the 14th-century canon who supported his archbishop against King Wenceslas IV, was thrown into it and drowned; suicides plunged at night into its depths.

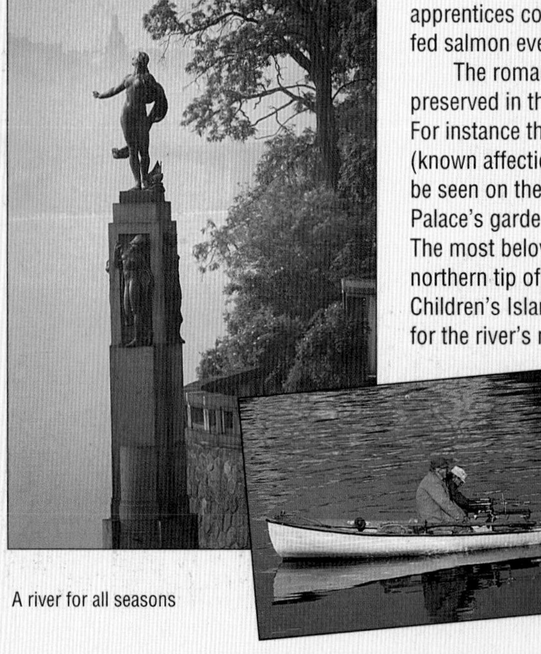

A river for all seasons

Nowadays, the once wild river, that regularly used to burst its banks, has been tamed with six dams along its 430km course from the Bohemian woods to its meeting with the Elbe just beyond Prague at Mělník. Winter skating is a thing of the past now that hydro-electric recycling has warmed up the water. In front of Charles Bridge it is a mere pond, with people splashing about in rowboats. Worse, the river has become increasingly unhealthy. The Vltava fish vaunted on Prague menus have to be caught upstream from where the dangerously polluted Berounka flows into the main river – a sad decline from the Middle Ages, when the city's apprentices complained that they were fed salmon every day.

The romantic image of the river is preserved in the city's memorials to it. For instance the Vltava water nymph (known affectionately as 'Terezka') may be seen on the wall of the Clam-Gallas Palace's gardens on Mariánské náměstí. The most beloved monument is on the northern tip of the Dětský ostrov (the Children's Island). From here wreaths for the river's many victims are thrown into the water in November. The city's dead are commemorated at All Souls (2 November), when a boat decked with black flags sets out on the dark waters.

A source of inspiration
and delight to
Czechs and visitors alike

The Belvedere Summer Palace, built by Ferdinand I, the first Habsburg king of Bohemia

BELVEDERE
It was between 1538 and 1563 that the Královská zahrada (Royal Gardens, see page 43) and a surpassingly elegant 'Lustschloss' (country seat) took shape on a patch of land some way to the northeast of Hradčany. This was the summer palace built by Ferdinand I for his wife, Anna Jagiello; the design and its garden setting show strong Italian influence.

The original architect of the 'Belvedere' (as the palace is known), Paolo della Stella, was responsible for the slim-columned arcade with its mythological reliefs. Bonifaz Wohlmut, the court architect who also built the Ball Game Court in the gardens, completed the ambitious project between 1552 and 1569. To him we owe the unusual copper roof, like the hull of an upturned ship, now a mellow green through oxidisation. Della Stella's arcade gives the palace its southern look and reminds one that it was a retreat for the court in the warm months. Rudolf II, Ferdinand's grandson, was particularly fond of it and encouraged his Danish astronomer, Tycho Brahe, to set up an observatory on the terrace. After Brahe's death, Johannes Kepler, his assistant, succeeded him as imperial mathematician and his *Laws of Planetary Motion*, published in 1609, were based on work done at the Belvedere.

The Swedes plundered the palace in 1648 and Joseph II turned it into a military laboratory. Only in the mid-19th

century was a proper restoration undertaken; when a cycle of historical painting dealing with the leitmotifs of Bohemian history was added to the first floor rooms.
Hradčany, Praha I. The Palace (now an exhibition hall) and the Royal Gardens open: May to September, Tuesday to Sunday 10am–6pm. Admission charge. Tram 22 to Belvedere.

BÝVALÝ BENEDIKTINSKÝ KLÁŠTER (Břevnov Monastery)

Legend has it that Duke Boleslav II and Bishop Adalbert of Prague were joint founders of Břevnov Monastery in 993, which would make it the oldest male convent in Bohemia. The site of the monastery was supposedly revealed to them in a dream. Such co-operation would be remarkable, if true, in that it pre-dates by only two years the massacre of Adalbert's (Slavník) family by their rivals for power, Boleslav's (Přemyslid) family.

The buildings

Apart from the crypt, there are hardly any Romanesque and Gothic architectural remains. The church, once dedicated to St Adalbert, was probably rededicated to St Margaret in the second half of the 14th century after her remains were transferred here in the 13th century. What we see today are the results of a complete rebuilding in baroque style between 1708 and 1745, by Christoph Dientzenhofer.

Dientzenhofer and his son, Kilián Ignác, turned Břevnov into one of the glories of Bohemian baroque. The Kostel sv Markéta (St Margaret's Church) is a masterwork of visionary architecture. Its marching series of diagonally protruding pillars support oval ceiling spaces covered with frescos. The tightly focused effect recalls Dientzenhofer's Church of St Nicholas in Malá Strana.

The monastery is undergoing restoration, but when it reopens there will be other important works to see, such as the frescos in the Prelates' Hall by Cosmas Asam. The Library has fine allegorical frescos by Felix Scheffler and the Refectory is notable for Bernhard Spinetti's stucco and Jan Kovář's manneristic painting on the ceiling. The Communists turned the monastery into an archive but the Benedictines are now back at Břevnov.
Markéta v Břevnově can be reached with trams 8 and 22, alighting at Břevnovský Klášter. Church open: 9am–6pm, but this could change because of building works. Other parts of the monastery should become accessible as soon as restoration is completed.

Spectacular ceiling frescos in the Břevnov Monastery

Cemeteries

MALOSTRANSKÝ HŘBITOV
(Lesser Quarter Cemetery)
The graveyard was opened in 1680 for plague victims, and continued in use until 1884.
Prague V. Trams 4, 7 and 9 to Plzeňská ulice, Bertramka.

OLŠANSKÉ HŘBITOV
(Olšany Cemetery)
This became one of the biggest burial grounds of Central Europe with over 100,000 graves. Among the remains of many celebrities lies student Jan Palach.
Prague III, Vinohradská třída. Trams 11 16 and 26. Metro: Flora.

STARÝ ŽIDOVSKÝ HŘBITOV
(Old Jewish Cemetery)
See pages 63 and 120.

VINOHRADSKÝ HVŘBITOV
(Vinohrady Cemetery)
Celebrated artists, actors, singers and writers are buried here, most notably journalist Egon Erwin Kisch.
Prague X, Vinohradská třída. Trams 11, 26.

VYŠEHRADSKÝ HŘBITOV
(Vyšehrad Cemetery)
Celebrities buried here include the novelists Karel Čapek and Jan Neruda, and composers Antonín Dvořák and Bedřich Smetana (see page 135).
Prague II, Vyšehrad. Metro station: Vyšehrad.

ŽIDOVSKÝ HŘBITOV
(Jewish Cemetery)
The grave of Franz Kafka is in section 21, row 14, at no 33.
Trams 11, 16 and 26. Metro: Želivského.

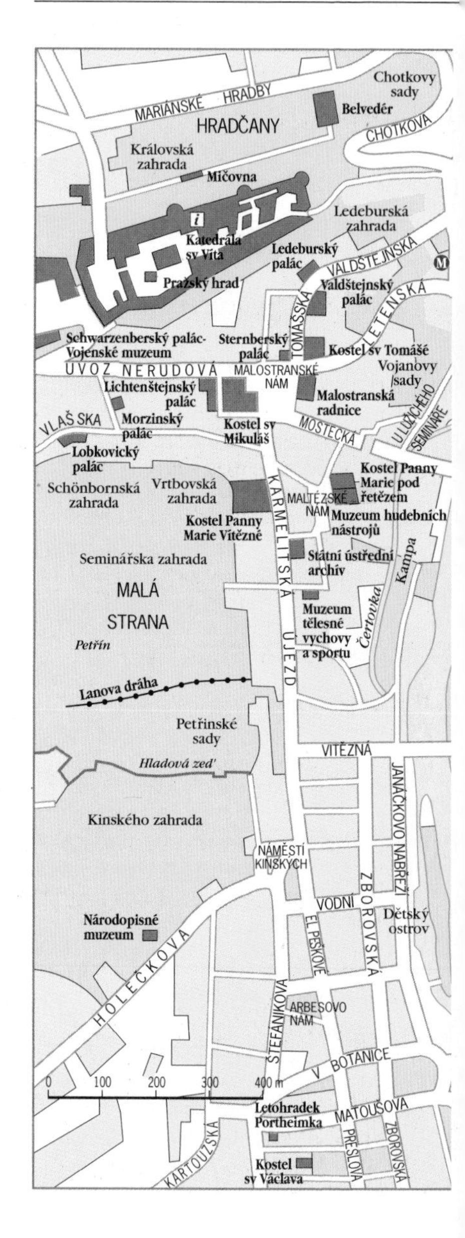

CENTRAL PRAGUE

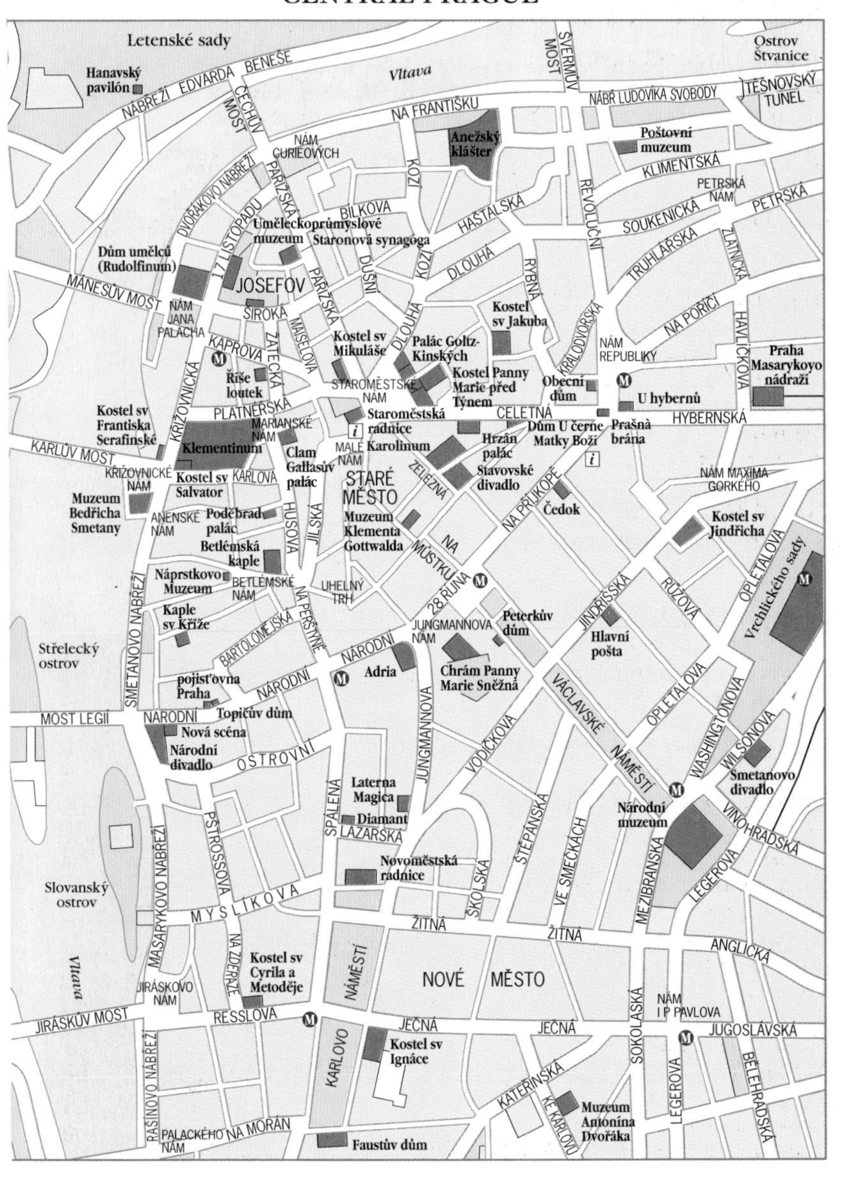

Churches

Prague boasts an astonishing wealth of ecclesiastical architecture: by the end of the 14th century there were 26 convents and monasteries in the city while scores of churches were built in the Middle Ages, the baroque period and the 19th century. This list of churches includes ones not covered under Hradčany, Staroměstské náměstí, and Malostranské náměstí. It excludes churches attached to convents where the latter have their own entries. Where access is limited, this has been noted.

BETLÉMSKÁ KAPLE (Bethlehem Chapel)
The present building is a modern replica of the Gothic trapezoidal chapel founded in 1391. It was rebuilt between 1536 and 1539, acquired by the Jesuits in 1661 and virtually demolished in 1786. The 14th-century Church authorities, who had agreed to the construction of a 'chapel', were faced with a building that could accommodate 3,000, a focus of church reform and a lasting spiritual centre of Hussitism.

In the adjoining preacher's house are exhibits relating to the Hussites and a reconstruction of a 15th-century domestic interior (see pages 118–9).
Staré Město. Betlémské náměstí. Restoration work may still be in progress, limiting access; otherwise it should be open between 9am and 5.45pm. Nearest metro stations: Můstek or Národní třída.

St James's Church in the Old Town, one of the loveliest churches in Prague

KOSTEL SVATÉHO CYRILA A METODĚJE (Church of SS Cyril and Methodius)

This somewhat forbidding church, completed by Kilián Dientzenhofer in 1740 was originally dedicated to St Charles Borromeo. When the Czech Orthodox Church took it over in 1935, it was rededicated to St Cyril and St Methodius. The crypt contains memorials of resistance fighters who took refuge there after assassinating the Nazi governor of Bohemia in 1942.

Nové Město. Resslova ulice. Access outside times of mass is difficult. Metro station: Karlovo náměstí.

KOSTEL SVATÉHO FRANTIŠKA SERAFINSKÉHO (Church of St Francis Seraphicus)

The richly decorated church (1689) is notable for its imposing cupola with V L Reiner's fresco of *The Last Judgement* (1722), and for its walls clad in Bohemian marble.

Staré Město, Křížovnické náměstí. Trams 17 and 18. Metro: Staroměstská.

KOSTEL SVATÉHO JAKUBA (Church of St James)

St James's is a baroque reconstruction of an earlier Gothic building (see page 117). The long glittering interior is furnished with 21 elegantly carved altars, above which are open galleries and a magnificent series of frescos (*The Life of the Virgin* and *The Adoration of the Trinity* by Franz Guido Voget). The altarpiece is Václav Reiner's version of *The Martyrdom of St James*. The impressive marble and sandstone tomb in the left-hand nave is of Count Vratislav of Mitrovic, a Bohemian Chancellor.

Staré Město, Malá Štupartská. Nearest metro: Náměstí Republiky or Můstek.

Jan Hus presides over Old Town Square

JAN HUS (c 1369–1415)

Born of peasant stock in Southern Bohemia, Hus had a meteoric career in the Church, ending up confessor to the Queen and Rector of Charles University. He was influenced by the English religious reformer, William Wycliffe (1320–84) and became the greatest preacher of his day. He collided with the Church on the issue of selling indulgences to finance papal wars and was summoned to the Council of Constance in 1415 where, despite the Emperor Sigismund's personal guarantee of safe conduct, he was convicted of heresy and burned at the stake.

In Prague there followed riots directed against the Church and its corrupt clerics. The Hussite wars that followed (1419–34) ended with an agreement at Basle between the Catholic Church and the more moderate wing of the Hussites. Only in 1965 did the Vatican overturn Hus's conviction for heresy.

The distinctive twin Gothic spires of St Peter and St Paul, Vyšehrad

KOSTEL SVATÉHO JANA NEPOMUCKÉHO NA SKALCE (Church of St John Nepomuk on the Rock)
Built on a precipitous site, this is one of Kilián Ignác Dientzenhofer's masterworks. A double flight of balustraded steps sweeps up to the twin-towered late baroque façade (1739). Inside, the ceiling fresco of *The Ascension of St John Nepomuk* by Karel Kovář is of exceptional quality, as is Jan Brokoff's altar statue of the saint, a model for the version later placed on Charles Bridge.
Nové Město, Vyšehradská třída. Access outside mass times is difficult. Metro and trams 3, 4, 14, 16, 18 and 24 to Karlovo náměstí.

KOSTEL NANEBEVZETÍ PANNY MARIE A KARLA VELIKÉHO (Church of the Assumption of Our Lady and Charlemagne)
Charles IV founded the monastery (part of the former Augustinian monastery known as Karlov) in 1350 and stipulated that the design for the church should follow Charlemagne's Imperial Chapel at Aachen, but some of the building's finest parts (such as the remarkable star vaulting) were completed by Bonifaz Wohlmut only in the 16th century. The pilgrims' steps on the south side are a baroque addition on the model of the Scala Santa in Rome. See the side galleries with theatrical sculptures by J J Schlansovský (*The Annunciation* and *Christ before Pilate*).
Nové Město, Ke Karlovu. Nearest metro station: I P Pavlova.

CHRÁM PANNY MARIE SNĚŽNÁ (Church of Our Lady of the Snows)
This building was planned as a Coronation Cathedral by Charles IV, but construction had got no further than the lofty choir before being interrupted by the Hussite rebellion. Jan Želivský, the radical reformer, was buried here after his execution in 1421. The angry demonstration that ended with the first defenestration of Prague began in Our Lady of the Snows on 30 July, 1419.
Nové Město, Jungmannova náměstí. Nearest metro stations: Můstek or Národní třída.

KOSTEL PANNY MARIE VÍTĚZNÉ (Church of St Mary the Victorious)
This was Prague's first baroque church (1613), ironically – since the baroque style is associated with militant Catholicism – commissioned by Lutherans. The Carmelites later rebuilt it. An object of veneration inside is the 'Bambino di Praga', a Spanish wax model of the infant Jesus with supposed miraculous powers.
Malá Strana, Karmelitská ulice. Trams 12 and 22 to Hellichova.

KOSTEL NEJSVĚTĚJŠÍHO SRDCE PÁNĚ (Church of the Sacred Heart)
Anyone with an interest in architecture should visit this church, a stunning example of Josip Plečnik's work, built in 1933 under the influence of the Viennese school of Otto Wagner. At the east end is a massive pedimental tower with a gigantic clock set in it like a rose window – a foretaste of contemporary Post-Modernism. The interior, though functionalist, has nobility and grace. Look for the stylised wooden statues of Bohemian saints by D Pesan, who also sculpted the monumental gilded Christ.
Vinohrady, Náměstí Jiriho z Poděbrad (metro station on the square).

Christ Child of Prague in St Mary the Victorious

The Church of the Sacred Heart in Vinohrady is a modernist masterpiece

KOSTEL SVATÉHO TOMÁŠE (Church of St Thomas)
Kilián Ignác Dientzenhofer remodelled this church in baroque style in 1731. Traces can be seen of earlier rebuildings of the Romanesque original in Gothic and Renaissance style. Ceiling frescos by Václav Reiner depict the *Life of St Augustine* and (in the cupola) the *Legend of St Thomas*. Karel Škréta, one of the greatest of Prague's early baroque painters, did several of the altarpieces. Over the high altar are copies of *The Martyrdom of St Thomas* and *St Augustine*, commissioned for the church from Rubens (the originals are in the National Gallery).

Buried in the cloister is an English humanist poetess at the court of Rudolf II, Elizabeth Jane Weston, (or 'Vestonia' under her Latin *nom de plume*).
Mála Strana, Letenská ulice. Trams 12 and 22 to Malostranské náměstí. Metro station: Malostranská.

Cubist Architecture

In 1910 the appropriately named Czech painter, Bohumil Kubišta, wrote excitedly home from Paris informing his colleagues that Picasso and Braque were the artists of the future. There followed a vogue in Prague for Cubist painting and Cubist design in the applied arts and architecture. In 1911 the influential *Skupina výtvarných umělců* (Group of Fine Artists) was established, with its own journal, and overnight (or so it seemed) Cubist houses appeared, filled with Cubist furniture and Cubist household articles.

The rapid acceptance of Cubism (and Modernism generally) demonstrated the determination of the Czechs to overcome their historical status of a provincial backwater. Cubism was attractive to artists and intellectuals who thought of Czech culture as part of the European mainstream, yet it was natural that there should be something characteristically 'Czech' about the local variants of Modernism.

Cubist buildings by Josip Gočár (1880–1948), Josef Chochol (1880–1956) and Pavel Janák (1882–1956) recall Bohemian baroque in their harmonious proportions and geometrical play of surfaces. Indeed, one of the earliest Cubist buildings by Gočár was accepted by the general public because it harmonised with its baroque neighbours on Celetná ulice.

After the founding of the Czechoslovak Republic, Cubism developed into 'National Style' or 'Rondocubism', intended as an expression of 'Slav' identity, but with both Cubist and traditional features. The façade of Janák's Rondocubist Adria Palace, which mixes all three elements, brings to mind a futuristic Italian Renaissance palace.

Although abstract form is central in Cubism, the play of geometrical surfaces in Prague's Cubist buildings provides a sensuality far removed from the arid anonymity of much modern architecture. Some people are impressed by their boldness, others find them sinister; either way, they are impossible to ignore.

CUBIST BUILDINGS

In 1912 Josip Gočár built the Dům U Černé Matky Boží (Black Madonna House) on Celetná ulice. It was so-called because a baroque statue of the Madonna from the previous house on the site was retained on the façade. There are four storeys, the first three of which consist almost entirely of windows, so that the building seems to be all eyes and no face.

Josef Chochol built more Cubist houses than any of his colleagues, whose designs often remained on paper. On Neklanova ulice in Vyšehrad is perhaps his most famous one, an apartment block with wedge-shaped forms on the façade and an overhanging cornice. Chochol designed the nearby Kovařovičova Villa on Libušina ulice, where even the garden and railings are Cubist. On Rašínovo

Shades of Renaissance: Janák's Adria Palace

nábřeží is his three-family house (Rodinný trojdům), described as 'like a classical palace', perhaps because of its relief-covered pediment.

On Ulice Elišky Krásnohorské is Otakar Novotný's Rondocubist building, one of the first to have a façade enlivened by colour contrasts. The best known Rondocubist work is Pavel Janák's Adria Palace. Another vivid example is the Bank Legií by Gočár on Na poříčí (now the Ministry for Industry) – with a patriotic frieze by Otto Gutfreund on the wall. Emil Králíček's Diamant (Diamond House – 1912) on Spálená ulice offers a coruscating display of variations on the diamond form – not beautiful but undeniably arresting.

BUILDINGS

Josip Gočár: Dům U Cerné Matky Boží (Black Madonna House), Celetná ulice 19; metro: Náměstí Republiky. Bank Legií, Na poříčí 24; metro: Náměstí Republiky.

Josef Chochol: Neklanova ulice 30 and Libušina ulice 3; trams 7, 18 and 24 to Na slupi. Rašínovo nábřeží 6–10; trams 3, 7 and 17 to Výtoň.

Otakar Novotný: Ulice Elišky Krásnohorské 123: tram 17 to Pravnická fakulta.

Pavel Janák: Adria Palace, Jungmannova 31; metro: Národní třída or Můstek.

Emil Králíček: Diamond House, Spálená ulice 4; Metro: Národní třída.

Gardens and Parks

Garden culture has a long tradition in Prague and reached its apotheosis in the baroque period. By the mid-18th century gardens bejewelled the north slope of Petřín Hill, the slopes beneath Strahov and the southern reaches of Hradčany. In the 19th and 20th centuries many of the city's most attractive green spaces became municipal parks.

The listing below excludes Petřín and Stromovka (see pages 102, 133 and 146) and the gardens of the Lobkowicz and Wallenstein palaces (see pages 95, 96 and 128). The castle gardens are included in the description of Hradčany (see pages 47 and 52) and those of Letná on page 126.

BOTANICKÁ ZAHRADA (Botanical Garden)

Even in the 14th century there was a botanical garden in Prague, run by a Florentine apothecary and situated where the main Post Office is now. The present garden is charming, but a bit run down (see page 125).

Na slupi 16. Open: daily 10am–7pm. Tel: 29 51 67. Trams 18 and 24 to Na slupi. Note: a new botanical garden has been established in Troja at Nádvorní 134, tel: 6641 0667. Open: daily 9am–6pm.

CHOTKOVY SADY (Chotek Park)

Founded by Count Chotek in 1833 as the city's first public park, Chotek Park contains a bizarre monument to the poet Julius Zeyer (1844–1901).

Trams 18 and 22 to Chotkovy sady.

KINSKÝCH ZAHRADA (Kinsky Gardens)

Prince Kinsky employed Franz Höhnel to lay out this park as an English landscape garden in 1825 and it passed to the municipality in 1901. Kinsky Gardens contain a neo-classical villa, once used by Crown Prince Rudolf and the Archduke Franz Ferdinand. Other curiosities are an 18th-century wooden church transported from a village in the

The Royal Gardens on the slopes of Hradčany

Plants for sale in the Botanical Garden (Botanická Zahrada)

Carpatho-Ukraine in 1929, and a picturesque little campanile from Moravia.
Entrance on náměstí Kinských (trams 6, 9 and 12) or from Petřín through the medieval 'Hunger Wall'.

KRÁLOVSKÁ ZAHRADA (The Royal Gardens)

Laid out in 1534, these gardens formed the first Italian Renaissance garden of Central Europe. A report of 1650 marvels at the pomegranates, figs, lemons and limes grown here, and at the tigers, lions, lynxes and bears kept in the menagerie. The botanist Mathioli cultivated the first European tulips here in the 16th century. At the north end is the Belvedere (see page 32), and on the east side you will pass the sgraffitoed Ball Game Court designed by Bonifaz Wohlmut, later used as a ballroom.
Open: April to September, Tuesday to Sunday 10am–5.45pm. Closed Monday. Admission charge. Tram 22 to Pražský hrad.

PALACE GARDENS BELOW PRAGUE CASTLE

Three gardens in Malá Strana were restored in the 1950s and opened to the public. The entrance is via the Kolowrat-Černín Palace, whose garden is laid out with rococo stairways, terraces, fountains and ornamental pools.
Valdštejnská 12. Open in summer only (currently undergoing restoration). Metro station: Malostranská.

VOJANOVY SADY (Vojan Park)

Tucked away in the lower part of Malá Strana is Vojan Park, former convent gardens. To the left of the entrance is a statue of St John Nepomuk standing on a fish. St Elias chapel is built as an imitation of a stalactite cave and has ceiling frescos of episodes in the saint's life.
Entrance from U Lužsického semináře. Metro station: Malostranská.

VRTBOVSKÁ ZAHRADA (Vrtba Gardens)

The Vrtba, among the finest of Prague's baroque terraced gardens, are noted for Matthias Braun's statuary: an Atlas near the entrance, and Ceres and Bacchus at the lower end.
Karmelitská ulice 26. The gardens are closed for restoration. Trams 12 and 22 to Malostranské náměstí.

PRAGUE LIFESTYLE

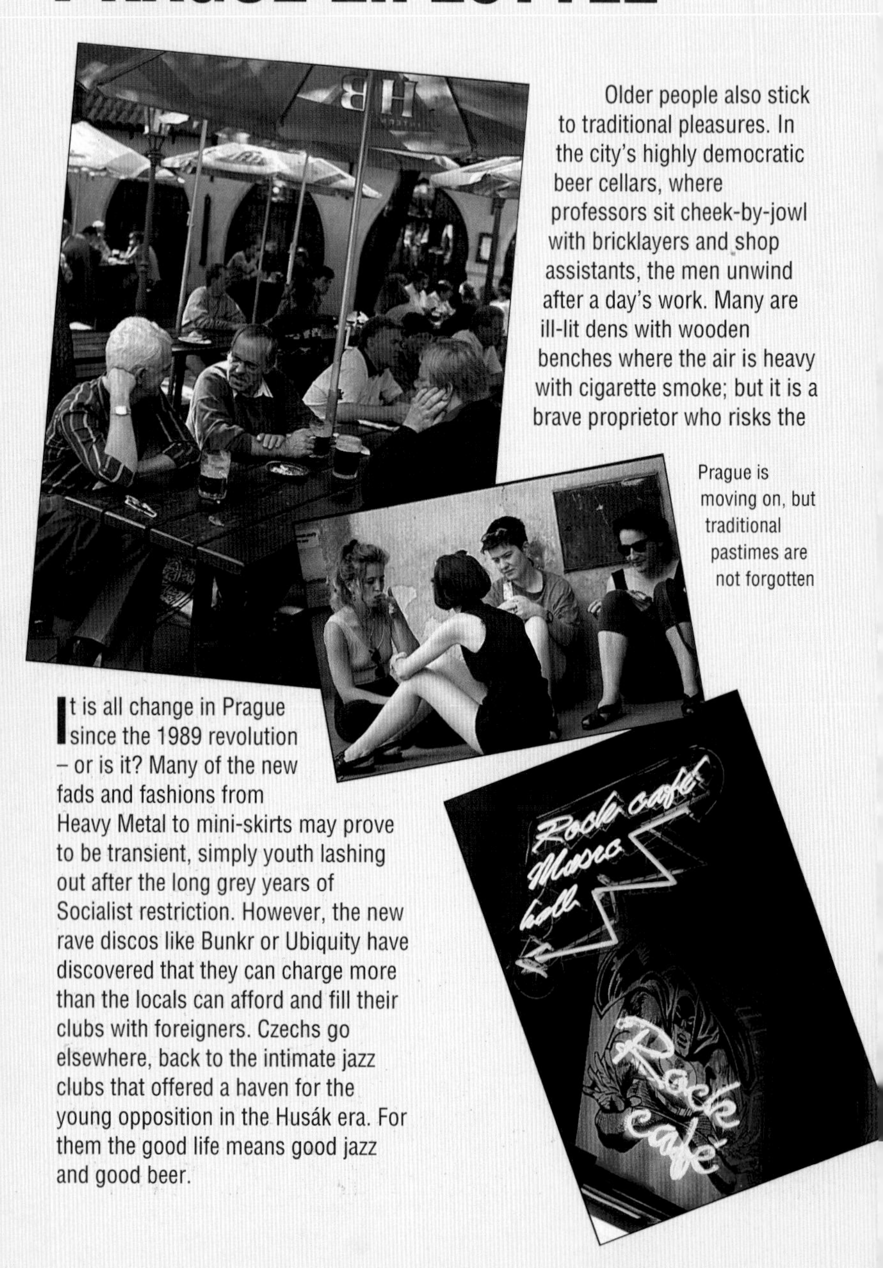

Prague is moving on, but traditional pastimes are not forgotten

It is all change in Prague since the 1989 revolution – or is it? Many of the new fads and fashions from Heavy Metal to mini-skirts may prove to be transient, simply youth lashing out after the long grey years of Socialist restriction. However, the new rave discos like Bunkr or Ubiquity have discovered that they can charge more than the locals can afford and fill their clubs with foreigners. Czechs go elsewhere, back to the intimate jazz clubs that offered a haven for the young opposition in the Husák era. For them the good life means good jazz and good beer.

Older people also stick to traditional pleasures. In the city's highly democratic beer cellars, where professors sit cheek-by-jowl with bricklayers and shop assistants, the men unwind after a day's work. Many are ill-lit dens with wooden benches where the air is heavy with cigarette smoke; but it is a brave proprietor who risks the wrath of his customers by tarting everything up, including the prices.

Meanwhile in the coffee houses the intellectuals still set the world to rights, while pensioners nurse Turkish coffees that must last them an hour. Coffee house culture has suffered from closures and changes of ownership, but several old stalwarts, such as Malostranská kavárna or the Paříž, soldier on.

Despite low salaries and high inflation, many Praguers own a dacha in the country. At weekends a procession of Škodas and Ladas heads out of the city for these often self-built second homes. Czechs are also zealous hikers. The areas around Prague are liberally signposted with routes for ramblers (the *Praha okolí* – environs of Prague – of the *Souboru turistických* map series could be useful for those who want to join in).

Walks and woods abound on the edge of the city

Hradčany

*T*he Pražský hrad (Prague Castle) lours over the Vltava just where an ancient trade route traversed the river. In the late 9th century it was fortified by Duke Bořivoj, and has been the spiritual and political focus of the Czech lands almost ever since. The term Hradčany is used to describe the whole area of the western hill above Malá Strana. It consists of Pražský hrad (the castle hill complex described here) and the former town of Hradčany to the west and north of it (see page 132), with Hradčanske náměstí (Hradčany Square, see pages 58–9) as its focal point.

After its diminutive and graceful beginnings in the Romanesque period (Bazilika sv Jirí – St George's Basilica and a rotunda Kostel sv Víta – Church of St Vitus), the architecture of Prague Castle reached its zenith under Charles IV (1346–78), when Matthew of Arras and Petr Parléř (Peter Parler) were at work on the great Gothic cathedral. In the late 14th century King Vladislav Jagiello commissioned the castle's finest secular architecture from Benedikt Ried – the jousting hall in the Royal Palace. The Habsburgs' architectural legacy, undistinguished apart from the Belvedere and the Royal Gardens, is depressingly pervasive. The castle area was clad in a barracks-like baroque classicism by Maria Theresa's architect, Nicolo Pacassi. The last architect of distinction to work on the buildings and gardens was Josip Plečnik, commissioned in the 1920s by Czechoslovakia's first president, Tomáš Masaryk, to remodel some of the baroque features and the bastions.

Standing proud at the main gate, Prague Castle

The Hrad lies at the heart of Czech consciousness and of Central European culture and history. It has seen imperial grandeur, and played host to humanist scholars, alchemists and occultists under the eccentric Rudolf II. It has been likened to Kafka's fictional castle under the Communists' rule of darkness; now it is the ceremonial residence of a democratic head of state.

THE BUILDINGS

PRVNÍ NÁDVOŘÍ (First Courtyard)

The tall wrought-iron gates of the entrance are topped by copies of battling Titans by the baroque sculptor, Ignaz Platzer. There is a changing of the guard at noon on Sundays (the accompanying fanfare played from the first floor windows was composed by rock star Michal Kocáb, who subsequently became an MP).

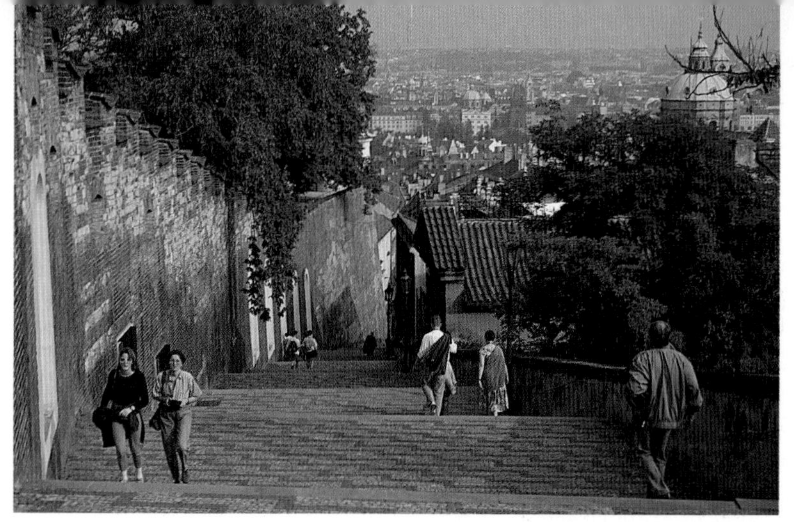

Count the steps to Hradčany, which rises above Malá Strana

Zahrada Na Baště (Garden on the Bastion)

The entrance to the left of the main gateway leads to the Bastion Garden, remodelled by Josip Plečnik in 1927. It has two levels connected by a circular stairway. The neo-classical pavilion is by Plečnik's successor as castle architect, Otto Rottmayer.

DRUHÉ NÁDVOŘÍ (Second Courtyard)

This is entered through the early baroque Matyasora brana (Matthias Arch), originally free-standing, later incorporated into Pacassi's rebuilding plan. In the north corner are the Rudofova galerie (Rudolf Gallery) and the Spanelsky sál (Spanish Hall), whose sumptuous neo-baroque interiors are not normally open to the public. From the northern exit of the castle, access is gained to Obrazárna Pražského hradu (the Castle Gallery, open: Tuesday to Sunday 9am–5 pm; admission charge), containing the remnants of Rudolf II's art collection. This once consisted of 3,000 pictures and 2,500 sculptures, but the Swedes looted most of it in 1648. Joseph II sold off much of the rest (Titian's *Leda and the Swan* was listed in the inventory as *Nude being bitten by an angry goose*). Among the remains is a bust of Rudolf by Adrian de Vries and works by Titian, Guido Reni and Rubens.

Kaple Svatého Kříže (Chapel of the Holy Cross)

The building in the southeast corner is Anselmo Lurago's Chapel of the Holy Cross, given its neo-classical aspect when it was altered in the 19th century. Formerly the Treasury, it has become a somewhat dreary art gallery.

> Pražský hrad, 119 08 Praha 1, Hradčany. The courtyards and streets of Castle Hill are open until late in the evening. The gardens are accessible between 9am and 5pm. All collections are closed on Mondays.
> Information Office: Vikarská 37.
> Open: Tuesday to Sunday 9am–5pm.
> Tel: 3337–1111. Metro (Line A): Malostranská. Tram 22 to Staroměstské náměstí, Malostranská or Pražský hrad.

KATEDRÁLA SVATÉHO VÍTA (St Vitus's Cathedral)

The earliest church on the site of St Vitus was a small Romanesque rotunda, founded by Duke (later Saint) Wenceslas around 925. The choice of St Vitus as the church's dedicatee may have been connected with the fact that the similarly sounding pagan god 'Svantovit' was previously worshipped here. Wenceslas also got a valuable propaganda boost for the new sanctuary in the form of St Vitus's arm, donated by the King of Saxony. Four centuries later Charles IV, who collected relics as lesser men collect postage stamps, was able to secure the rest of St Vitus for his treasury. This was a major coup, for the cult of this highly prophylactic saint (invoked against epilepsy and 'St Vitus's dance' or chorea) was hugely popular in the Middle Ages.

In 1039 the bones of the most important local martyr, St Adalbert, were also placed in the rotunda. It thus became a much frequented pilgrimage shrine; so much so that a larger church had to be built in 1060, a towered Romanesque basilica dedicated to Saints Vitus, Adalbert and Wenceslas, whose tombs were inside.

Some remains of the original rotunda and basilica came to light in the 1920s as the cathedral was undergoing its last phase of building. They can be seen below the cathedral floor.

Open: Tuesday to Sunday 9am–5pm. Crypt open: 9am–4.45pm. Admission charge.

Bronze sculptures adorning St Vitus's

The building of the Gothic church

When Prague became an archbishopric in 1344 the future Charles IV summoned Matthew of Arras from Avignon to build a cathedral. His design was in the severe Gothic style of contemporary church architecture in France. He had completed eight chapels by his death in 1352, then in 1356 Petr Parléř, a Swabian, took over the work and it is to his genius that we owe the altogether overwhelming power and beauty of St Vitus's architecture. In particular it is his vision that produced the striking contrast between the complex ornateness of the exterior walls and the calm immensity of the interior.

By 1366 Parléř had completed the Chapel of St Wenceslas. In 1385 the choir was consecrated, at which time a 'temporary' wall was erected between it and the nave (which was still under construction). Little did the optimistic builders imagine that this wall would still be in place nearly five centuries later!

Parléř and his workshop began the triforium with its busts of dynastic and other figures and laid the foundation stone for the great South Tower in 1392. After he died in 1399 his sons Wenceslas and John worked on for 20 years.

The Hussite wars (see page 37) brought building to a halt, but there was a final phase of Gothic construction under King Vladislav Jagiello, whose architect was responsible for the unusual Royal Oratory, built in the 1480s.

In 1843 the Association for the Completion of St Vitus Cathedral was set up to raise funds. In the second half of the 19th century Josef Mocker

The imposing façade of St Vitus' Cathedral, Prague's largest church

carried through plans that respected the spirit of Parlér's Gothic design, but also managed to salvage important Renaissance and baroque features that had been added over the years. Work continued into the 20th century and it was not until the millenary of the murder of St Wenceslas (12 May 1929) that the entire church was considered to be complete and finally dedicated.

A TOUR OF ST VITUS CATHEDRAL

On entering the third courtyard of the castle you are confronted with the cathedral's main entrance in its neo-Gothic western section, but walk round to the south side to admire Petr Parléř's fine architecture. He built the Gothic part of the 96m-high bell-tower, which is topped by a Renaissance gallery and Pacassi's baroque spire. To the right is the triple-arched Golden Portal and higher up is a remarkable openwork stairway, a good example of the unconventional boldness of Parléř's design.

People have worshipped in churches on this site for a thousand years

Kaple sv Václav (St Wenceslas Chapel)

Immediately on your right as you enter through the south portal is the chapel built to hold the remains of Bohemia's most beloved patron saint. Some 1,370 precious stones are set in its lower walls, perhaps signifying the year it was completed. Frescos depict the New Heavenly Jerusalem, the Life of Wenceslas and the Passion of Our Lord; there is also a statue of the saint by Parléř's nephew, Heinrich, and the (much restored) Wenceslas tomb. The Crown Jewels are kept above the chapel but are seldom on view to the public.

The Royal Oratory

Beyond the two chapels to the east of the St Wenceslas chapel is the Royal Oratory, a hanging vault with intricate Gothic decoration resembling the branches of a tree. Designed for Vladislav Jagiello, it was connected with the king's bedroom in the palace by a covered gangway. The baroque figure on the left-hand edge represents a miner from the silver mines at Kutná Hora (see pages 140–1).

The Tomb of St John Nepomuk

Two chapels further is the baroque tomb of St John Nepomuk by Fischer von Erlach the Younger (1736). The cherub on the lid is pointing to the saint's tongue, a reference to the Jesuit claim that this part of his remains had never decayed. (According to legend the saint had refused to betray the secrets of the Queen's confession to Wenceslas IV.)

Panel reliefs

In the north ambulatory, opposite Parleř's Old Sacristy, is a finely carved oak panel with a relief depicting the

The tomb of St John Nepomuk

Elector Frederick's hasty retreat from Prague after the defeat of the Protestant Bohemians at the White Mountain. A complementary panel in the south ambulatory shows Hussites plundering the cathedral in 1619.

Other sights of interest

In the centre of the cathedral is the Royal Mausoleum, a fine Renaissance work with reliefs of Bohemian kings. Next to it is the entrance to the crypt and archaeological remains of the two earlier churches on this site. The third chapel from the west end on the north wall contains Alfons Mucha's stained-glass window of Saints Cyril and Methodius. Next to it is František Bílek's remarkable *Crucifixion* (1899).

Finally, it is worth taking binoculars to see the 21 sandstone busts by the Parléř school high up in the Triforium.

ST WENCESLAS

The door of the chapel has a lion's-head knocker to which, it is claimed, the dying Wenceslas clung when struck down by his brother Boleslav. The murder took place 20km outside Prague at Stará Boleslav. The motives for the killing appear to have been dynastic and political rather than religious; nevertheless Wenceslas was treated as a martyr. His remains were subsequently transferred to St Vitus, apparently by a now repentant Boleslav. J M Neale's famous carol (which promotes Duke Wenceslas to king) is unhistorical, although the image of him as a just and merciful ruler may have had some substance.

HRADČANY–PRAŽSKÝ HRAD

JELENÍ
U PRAŠNÉHO MOSTU
MARIÁNSKÉ
Luí dvůr
Jízdárna Pražského hradu
Královská
Míčovna
Jelení
Šternberský palác-Národní galérie
KANOVNICKÁ
Martinický palác
PRAŠNÝ MOST
Obrazárna Pražského hradu
Španělský sál
Mihulka
Zahrada Na baště
Toskánský palác
Arcibiskupský palác
druhé nádvoří
VIKÁŘSKÁ
Staré proboštství
katedrála sv Víta
NÁM U SV JIŘÍ
Kohlova kašna
Kostel sv Benedikta
LORETÁNSKÁ
HRADČANSKÉ NÁMĚSTÍ
První nádoří
Kapel sv Kříže
Socha sv Jiří
Starý Královský palác
Tretí nádvoří
Vladislavský sál
Matyasova brána
RADNICKÉ SCHODY
Schwarzenberský palác-Vojenské muzeum
KE HRADU
ZÁMECKÉ SCHODY
Rajská zahrada
Zahrada
Hudební pavilón

THIRD COURTYARD

Zahrada Na Valech (Rampart Garden)

On the south side of the courtyard a flight of steps gives access to the Rampart Garden with a viewing terrace by Josip Plečnik (1924). At the west end it merges with the Rajská zahrada (Paradise Garden), changed by Plečnik, but originally laid out in 1562.

The vaulted ceiling of the Vladislav Hall

Starý Královský Palác (Old Royal Palace)

On the southeast side of the courtyard is the entrance to the Royal Palace of the Bohemian rulers between the 11th and 16th centuries (open: Tuesday to Sunday 9am–5pm; admission charge). The top part was chiefly built by Benedikt Ried for King Vladislav Jagiello in the 15th century. Ried was responsible for the Vladislavský sál (Vladislav Hall), entered from the antechamber. On the way, note to your left the Green Room, once the Supreme Court, and Vladislav's Bedchamber. Between 1493 and 1502

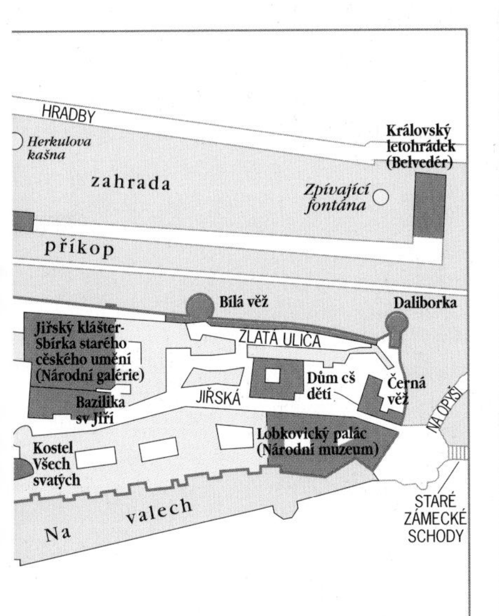

this great jousting and banqueting hall (16m wide, 62m long, 13m high) was constructed in the late Gothic style. Light streams in from the huge Renaissance windows on either side. In Rudolf II's time the space was used as a bazaar and since 1918 the presidents of the republic have been sworn in here. The door in the southwest corner gives access to the Bohemian Chancellery in the Louis Wing of the palace, scene of the third defenestration of Prague in 1618. After returning to the Vladislav Hall walk to the far end where there is a viewing platform offering fine views over Prague. East of the hall is the rather dreary Kostel Všech svatých (All Saints Chapel).

More interesting is the Hall of the Diet in the northeast corner, the work of Bonifaz Wohlmut (1563). Portraits of

The Romanesque splendour of St George's Basilica, east of St Vitus

18th- and 19th-century Habsburgs adorn the walls.

To the left is the Riders' Stairway, up which the knights rode to their tournaments in the great hall. At the bottom are the earlier Gothic and Romanesque parts of the palace, where copies of some of the busts on the triforium of St Vitus are displayed, together with models of the historical development of the castle.

Outside St George's Basilica

JIŘSKÉ NÁMĚSTÍ

Leaving the Royal Palace by the Riders Staircase you enter Jiřské náměstí (St George's Square).

Klášter Sv Jiří (Convent of St George)

In the northeast corner of the square is the Convent of St George. Its foundation in 973 marked the elevation of Prague to a bishopric and its first abbess was Mlada, sister of the then ruler, Boleslav II. It remained a convent for Benedictine nuns until Joseph II turned it into a barracks in 1782. Since 1972 it has housed the National Gallery's Collection of Old Bohemian Art (see page 90). *Jiřske náměstí 33. Tel: 2451 0695. Open: Tuesday to Sunday 10am–6pm. Admission charge.*

Bazilika Sv Jiří (Basilica of St George)

Adjoining the convent to the south is the Basilica of St George. Restoration at the turn of the century and in the 1960s has made this Prague's best preserved Romanesque building, although it has inevitably lost something of its ancient lustre. Inside there are three aisles, the middle one a great barn-like hall with a flat wooden ceiling. At the east end a triumphal arch frames a raised altar approached by a double flight of baroque steps. In front of the steps are the tombs of Boleslav II and Duke Vratislav (the founder of the church) and between them is the entrance to the 12th-century crypt. The church's origins go back to about 920; the last alterations were in 1680, when the chapel of St John Nepomuk was built at the southwest end and the somewhat incongruous baroque façade added.

Inside St George's Basilica

At the southeast end is the chapel dedicated to St Ludmila. She became Bohemia's first martyr. It contains her tomb, designed by Petr Parléř, and frescos depicting her life and martyrdom by J V Hellich (1858).
Open: Tuesday to Sunday 9am–5pm. Admission charge.

Lobkowický Palác (Lobkowicz Palace)
South of the basilica at Jiřská ulice no 1 is the former seat of the Lobkowiczes, who were fervent protagonists of the Counter-Reformation. In 1618, the redoubtable Polyxena Lobkowicz took in the two councillors ejected by Protestant nobles from a Chancellery window, and refused entry to their pursuers. Two floors of the palace are given over to a museum of Bohemian history.
Jiřská ulice 1. Tel: 53 73 06. Museum of the History of Bohemia open: Tuesday to Sunday 9am–5pm. Admission charge.

Over the entrance to Lobkowický Palác

BOHEMIA'S FIRST MARTYR
Ludmila and her husband, Duke Bořivoj 1 (870–94, had been baptised by Methodius shortly before the missionary's death in 885. She thus played a key role in the Christianisation of the Czech lands, particularly through the influence she exerted over her grandson, the future Duke Wenceslas. This was bitterly resented by Drahomíra, Wenceslas's mother, who became regent on the young duke's accession at the age of 13. Ludmila was forced to retire from Prague to Tetín near Karlštejn. It was here that Viking mercenaries sent by Drahomíra strangled her with her own veil as she lay defenceless in her bedchamber on the night of 15 September 921.

Drahomíra soon realised that she had miscalculated. Pilgrims came from far and wide to Ludmila's grave. In desperation, Drahomíra ordered a church to be built over it and dedicated it to the Archangel Michael, so that the constant miracles taking place should be ascribed to him, not Ludmila. The people reacted by honouring Ludmila even more. In 925 Wenceslas had the remains of his grandmother brought to Prague and ceremonially deposited in St George's Basilica. They proved to be miraculously undecayed and even gave off an agreeable odour, two indisputable indications of saintliness.

Zlatá Uliča (Golden Lane)

At the north end of the castle complex (approached from the rear of St George's Convent) is Golden Lane. The emperor's 24 scarlet-uniformed gate-keepers originally lived in its little houses, pursuing various crafts and trades to eke out their miserable wages. Later goldsmiths and other artisans arrived. The Renaissance house backing on to the south side of the lane belonged to the burgrave (the king's deputy).

Number 22 was occupied by Franz Kafka in 1916–17; the Kafka society now owns it and plans to turn it into a bookshop. Nobel Prize-winning poet, Jaroslav Seifert, also lived in Golden Lane.

The diminutive houses of Golden Lane

Prašna věž/Mihulka (Mihulka or Powder Tower)

At the western end of Golden Lane is the Powder Tower, where Rudolf II's alchemists laboured to find the formula for making gold. Among those who worked on this doomed project were Englishmen John Dee (formerly employed by Elizabeth I) and Edward Kelley. They were both later disgraced for failing to deliver the goods. Kelley was to end up a prisoner in the castle of Křivoklát (see page 139). On display are the alchemists' equipment, together with some fine 16th-century furniture and works of art.

The appellation 'Mihulka' apparently refers to the lampreys (*mihule*) bred here for the royal kitchens.

Open: Tuesday to Sunday 9am–5pm. Admission charge.

Daliborka věž (Dalibor Tower)

This tower, at the east end of the lane, takes its name from the nobleman imprisoned there in 1498 on suspicion of aiding and abetting a peasants' revolt that had broken out near Leitmeritz. Legend says he passed the days before his execution playing the violin so movingly that passers-by filled the basket he let down from his window with victuals and money. Bedřich Smetana used the story for one of his best-known operas, *Dalibor*.

Open: Tuesday to Sunday 9am–5pm.

Bílá věž (the White Tower) and Černá věž (the Black Tower)

The White Tower (midway along Golden Lane) and the Black Tower at the eastern tip of the castle were prisons, the latter for debtors. The legend goes that the 16th-century

Gateway to the Old Town, the 65m-high late Gothic Powder Tower

aristocrat Katharine Lažan died in the White Tower, accused of having murdered several young female servants in order to retain her beauty by washing in their maidenly blood.

Another inmate was Kaspar Rucký, Rudolf's corrupt treasurer. He used the silken cord that had held the keys to the treasury to strangle himself, thereby escaping a considerably worse fate.

Hradčanske Náměstí

(Hradčany Square)

The township of Hradčany dates back to 1320 and originally consisted of little more than the square itself. Following a devastating fire in 1541, most of the burghers' houses were pulled down by the Catholic nobility, who bought up large plots and built great palaces on them. In the centre of the square is Ferdinand Brokoff's Marian Column (1726), with representations of the Bohemian patron saints around its base. To the west of the Arcibiskupský palác (Archbishop's Palace) look out for a cast-iron candelabra gaslight dating to the 1860s (see pages 130 and 132).

SCHWARZENBERSKÝ PALÁC
(Schwarzenberg Palace)

At no 2 on the south side of the square the Schwarzenberg Palace has a number of striking Italianate features, including Lombardy cornices and Venetian-style sgraffiti. When Agostino Galli originally built it for the Lobkowiczes in 1563 seven existing houses on the site had to be demolished. Inside (on the second floor) are tempera frescos depicting scenes from Homer.

The Schwarzenbergs acquired the building only in 1719. A previous owner has entered history for having invited Tycho Brahe (the imperial mathematician) to a party in 1601. It proved to be the irascible astronomer's last outing: according to the story, as a result of over-indulgence at dinner his bladder burst on the way home.

The interior of the Schwarzenberg Palace can be seen by those visiting the Vojenské Muzeum (War Museum) located here. Open: Tuesday to Sunday 8.30am–5pm. Admission charge (see page 83).

ARCIBISKUPSKÝ PALÁC
(Archbishop's Palace)

Close to the castle's gates at no 16 is the Archbishop's Palace, boasting an elegant rococo façade by Johann Wirch (1764).

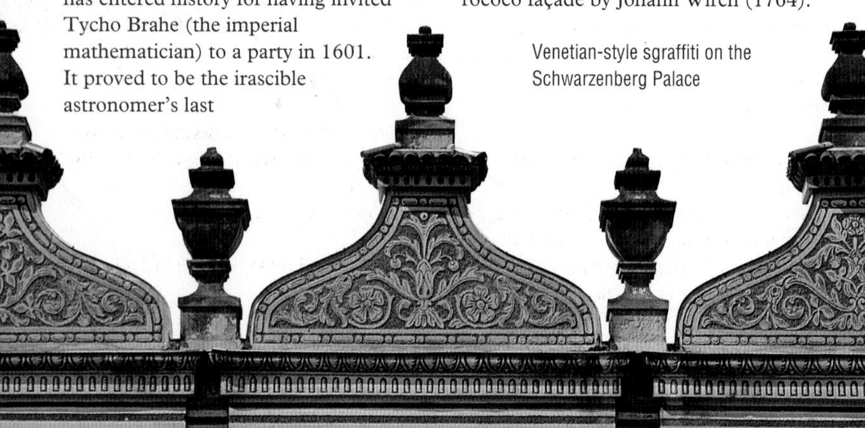

Venetian-style sgraffiti on the Schwarzenberg Palace

Traces of Jean-Baptiste Mathey's earlier baroque design (1676) may be seen in the entrance portals and the tympanum rising above the middle of the façade. With this first commission in Prague Mathey introduced many of the architectural principles of the Italian baroque imbibed during his period in Rome.

The original Renaissance palace on this site was presented by Ferdinand I to the first post-Hussite archbishop of Prague, whose residence thus moved closer to the centre of power on Pražský hrad. Bonifaz Wohlmut rebuilt it in 1564.

The interior is only open to the public once a year on Maundy Thursday between 9am and 5pm.

ŠTERNBERSKÝ PALÁC (Sternberg Palace)

Through the left-hand entrance arch of the Archbishop's Palace access is gained to the Sternberg Palace at no 15, a building of the high baroque designed by Giovanni Alliprandi. The Chinoiserie room on the second floor is notable, but the main reason for visiting the palace is the National Gallery's Collection of Old European Art (see page 91).

Open: Tuesday to Sunday 10am–6pm.

MARTINICKÝ PALÁC (Martinic Palace)

Restoration during 1971 brought to light sgraffiti depicting the story of Joseph and Potiphar on the impressive Renaissance façade of the Martinic Palace (no 8). Its owner from 1624 was Jaroslav Bořita of Martinitz, one of the councillors defenestrated from the Bohemian Chancellery in 1618 – and afterwards made a count for his troubles. Both he and his fellow-victim, Vilém Slavata, arranged to have their miraculous escape immortalised in bombastic sculptures for their palaces.

The palace is not open to the public.

The Archbishop's Palace, with its magnificent façade, is a masterpiece of rococo

TOSKÁNSKÝ PALÁC (Tuscany Palace)

This huge and rather sombre palace, at the west end of the square (no 5), was built by Jean-Baptiste Mathey in 1689–91 and owned by the Dukes of Tuscany between 1718 and 1918. It now belongs to the Czech Foreign Ministry.

Josefov

The Jews started arriving in Prague in the 10th century and settled on both sides of the Vltava. Their first eviction was at the hands of Otakar II, who needed their land for his new town of Malá Strana on the west bank. From that time until the deportations to concentration camps under the Nazis, theirs was a history of recurrent victimisation and violence.

The two worst pogroms of the Middle Ages were in 1086, when the Crusaders indulged in an orgy of Jew-killing, and 1389, when 3,000 Jews were massacred over Easter. In 1745 Maria Theresa expelled the Jewish population from Prague, but had to allow them back soon afterwards under pressure from commercial interests. Joseph II's Edict of Tolerance in 1781 improved their lot, but with the rise of Czech nationalism in the 19th century many German-speaking Jews found themselves on the wrong side of the cultural divide. Nevertheless, in the first half of the 20th century, German Jewish literary culture flourished, producing a string of major writers including Franz Kafka, Max Brod, Egon Erwin Kisch and Franz Werfel.

All that ended with the Nazis, who killed 80,000 of the 90,000 Jews who remained in Bohemia after the invasion. A grotesque footnote to genocide was Hitler's decision to found a Jewish Museum in Prague, which he designated an 'Exotic Museum of an Extinct Race'. The Communists took this over, but chose to exploit it for self-glorification.

Josefov is reached by metro to Staroměstská or Tram 17. *See walk on page 120.*

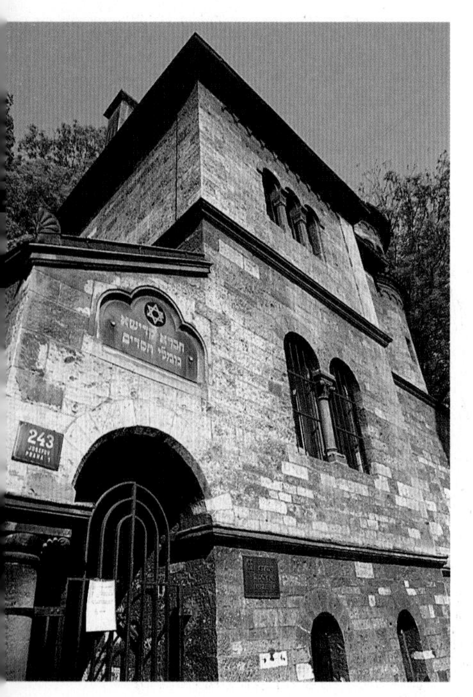

Exhibitions are held in the Ceremonial Hall

The ghetto

The ghetto was built in the 13th century in accordance with the church's view that Jewish dwellings should be kept separate from those of Christians. From time to time ordinances regarding clothing – designed to mark out and humiliate Jews – were promulgated. Under Vratislav II they had to wear yellow cloaks; later it was bizarre hats or yellow circles.

Under Rudolf II, whose financial adviser was the Jew Mordechai Maisl, the Jewish community achieved a greater degree of autonomy. The Emperor's obsession with the occult also engendered an interest in cabbalistic lore and learning (this was the age of Rabbi Jehuda Löw, credited with creating the 'Golem' – see page 63).

In 1784 Jewish residence restrictions were abolished following Joseph II's Edict, but by the end of the 19th century the ghetto area had become an insanitary slum and red light district. In 1893 many buildings were razed to make way for a modern quarter.

Art nouveau doorway

STÁTNÍ ŽIDOVSÉ MUZEUM
(State Jewish Museum)

A single ticket covers access to all the sights of the State Jewish Museum listed below. Tickets may be obtained from the Klausová Synagóga on U starého hřbitova. Open: April to October, 9am–5pm; November to March, Sunday to Friday 9am to 4.30pm. Last tickets are issued half an hour before closing time and most monuments close over lunch, noon to 1pm. Office: Jáchymovo 3. Tel: 2481 0099.

KLAUSOVÁ SYNAGÓGA (Klausen Synagogue)

This 17th-century building houses a permanent exhibition of prints and early manuscripts. Rabbi Löw is said to have had his school here.

MAISELOVÁ SYNAGÓGA (Maisel Synagogue)

Mordechai Maisel gave 12,000 denars for the construction of the synagogue in 1590. It now contains a fine silver collection with objects made by Renaissance masters from Augsburg and Nuremberg and artefacts of the Bohemian and Moravian baroque.
Maiselova ulice.

OBŘADNÍ SÍŇ (Former Ceremonial Hall of the Gravediggers' Company)

This neo-Romanesque building at the Old Jewish Cemetery entrance exhibits drawings and paintings by children in Terezín concentration camp. Some 15,000 Jewish children were held there before being deported to Auschwitz, where the majority perished.
U starého hřbitova.

Universal Jewish symbol

PINKASOVA SYNAGÓGA
(Pinkas Synagogue)

A Rabbi Pinkas founded the earliest synagogue on this site in 1479; the Gothic vault of the interior was built in 1535 and the women's gallery added a century later. On its walls are inscribed the names of 77,297 Bohemian and Moravian victims of the holocaust.
Široká ulice.

ŠPANĚLSKÁ SYNAGÓGA
(Spanish Synagogue)

The Spanish Synagogue is an Alhambra-like building with neo-Renaissance features, the work of Ignaz Ullmann (1864). Sephardic Jews expelled from Spain at the end of the 15th century built a previous synagogue on this site.
Dušní 12.

STARONOVÁ SYNAGÓGA
(Old-New Synagogue)

Built between 1270 and 1280 this is the most impressive of Josefov's synagogues. The design incorporates an unusual five-ribbed vaulting inside. In the vestibule are chests placed for the collection of taxes. Over the entrance to the hall is a relief of a vine with 12 bunches of grapes, supposedly representing the 12 tribes of Israel. Inside are the *almemor* (a pulpit behind a Gothic lattice) and a shrine for the *Torah* (a parchment scroll of the Pentateuch). A magnificent statue of Moses by František Bílek stands next to the synagogue in a small park.
Červená ulice.

The perversely named Old-New Synagogue is Josefov's loveliest attraction

People have worshipped here for more than 600 years

STARÝ ŽIDOVSKÝ HŘBITOV
(Old Jewish Cemetery)

Over 550 years the remains of some 20,000 people have been bundled into 12 layers of graves in the Old Jewish Cemetery (see page 120).

The earliest grave is that of poet Avigdor Kara who survived and chronicled the 1389 pogrom. Also interred here is Mordechai Maisel, who amassed 17,000 *gulden* from trade monopolies under Rudolf II and was the ghetto's greatest benefactor. Rabbi Löw's tomb is still sprinkled with pebbles, prayers and even money – evidently the old cabbalist still exerts a spell.

The headstones are works of art in themselves, with carved symbols indicating the deceased's profession, or the family name. It is a strange and haunting place, evoking both the forlorn dignity and astonishing resilience of the Jewish community.
U starého hřbitova.

VYSOKÁ SYNAGÓGA
(High Synagogue)

In the High Synagogue's fine Renaissance interior there is an exhibition of Jewish textiles, many gathered during the war for the Jewish museum projected by the Nazis.
Červená 4.

ŽIDOVSKÁ RADNICE
(Jewish Town Hall)

The rococo aspect of the Jewish Town Hall is the result of alterations by Josef Schlesinger in 1765, when it also acquired its backwards reading Hebraic clock. The hall is still the community centre for Prague's 1,000 Jews and also boasts a kosher restaurant.

The Nazis employed Jewish scholars here to collect material for the planned 'Museum of an Extinct Race' (see page 60) until early 1945, when the last of them were despatched to the camps.
Maiselova 18.

THE GOLEM LEGEND

In 1580 when the perennial accusations of ritual murders and other crimes were being levelled at the Jews, Rabbi Löw decided that the ghetto needed reassurance and protection. Using his cabbalistic knowledge (so the story goes) he was able to create a humanoid from the mud of the Vltava, the 'Golem' (Hebrew for 'unformed matter'). Its job was to act as servant and bodyguard to the community: when not required, it was switched off by placing a secret formula (*shem*) in its mouth. Like many a later Frankenstein's monster, the Golem finally escaped the control of its creator and ran amok in the rabbi's house (Löw had forgotten to put the *shem* in his mouth). For this rebellion the humanoid was promptly turned back into a lump of clay and deposited in the attic of the Old-New Synagogue.

Top left: Hunger Wall
Above: Number 7 Na Poříčí
Left: Kafka's grave in the Jewish Cemetery
Right: 22 Golden Lane

Kafka and Prague

Largely ignored in the author's lifetime, burned by the Nazis and suppressed by the Communists, Kafka's work is at last being celebrated in the city of his birth. Franz Kafka (1883–1924) wrote *The Trial* (1925) and *The Castle* (1926) and a body of short stories creating a fictional world both absurd and menacing. Gregor Samsa, the hero of *Metamorphosis*, awakes one morning to find he has turned into a giant beetle; the hero of *The Trial* is suddenly arrested, but never discovers why.

Prague, the city that Franz Werfel claimed 'has no reality', looms over Kafka's stories as it did over his life. Pražský hrad inspired *The Castle*, and the Hladová zed (Hunger Wall – see page 133) on Petřín Hill his story *The Great Wall of China*. Echoes of the Prague ghetto reverberate through the fantastical world of his imagination. The bumbling bureaucracy of the Habsburg administration in the 19th century supplied Kafka with numerous ideas. In retrospect his vision seems to be a premonition of what life for the people of Prague would be like under Nazism and Stalinism, but Kafka did not live long enough to experience it.

Devotees of Kafka's work can follow his trail through Prague – to his one-time workplace (Na poříčí 7), to the house he rented in Golden Lane on Castle Hill (no 22), and to his grave in the Jewish Cemetery at Strašnice (Metro line A to Želivského). The city's beauty is all around us, but reading Kafka makes us aware of another, more menacing, presence behind the beautiful façade. Kafka had this in mind, perhaps, when he wrote in his diaries: 'Prague is a dear mother with sharp claws: she never lets go of you.'

U HYBERNŮ (Hibernian House)
The name of this building recalls the Irish Franciscans who occupied a baroque monastery and church on this site from 1629 until the dissolution of their establishment under Joseph II in 1786. After briefly being used as a theatre, the former church was reconstructed in neo-classical (or 'Empire') style in 1811 to plans by Johann Fischer. The Imperial Government used it as the Central Customs Office for Prague (hence the imperial double-headed eagle in the side-wall tympanum). After World War II it became an exhibition hall. Fischer's design, modelled on the Old Mint in Berlin, is the most completely realised example of Empire style in Bohemia. *Náměstí Republiky 3. Metro: Náměstí Republiky. Open during exhibitions.*

The Empire-style U Hybernů is a venue for exhibitions and cultural events

KAROLINUM (Carolinum/Charles University)
Charles IV founded Europe's 35th university in Prague on 7 April 1348. The Charles University, or Karolinum, was the first such foundation in Central Europe, and its professors and students were entitled to teach and study at any other school sanctioned by the Catholic church.

This all came into question from 1409 onwards, after the Hussite faction in Prague pressured Wenceslas IV into issuing the Decree of Kutná Hora, whereby the Czechs gained the upper hand in the university administration. There was a mass exodus of non-Bohemians (leading to the founding of Leipzig University), and Jan Hus became rector of a school increasingly regarded elsewhere in Europe as a nest of heretics. Eventually, in 1412, the Catholic interest rallied its forces and succeeded in having Hus ejected.

After the Battle of the White Mountain in 1620 the Jesuits (who had had their own 'Clementinum' since the mid-16th century) were allowed to annex the Carolinum.

The buildings
Some remains of the original buildings have either been laid bare during modern restoration work or were 're-Gothicised' by Joseph Mocker in the late 19th century. The finest surviving feature is the lovely oriel window (circa 1370) overlooking Ovocný trh (the fruit market). Much of the rest was reworked in baroque style in the 18th century.

The 17th-century Assembly Hall on the first floor with a tapestry depicting

Charles IV and paintings on the organ loft by V Sychra is unfortunately not always accessible. However, you can visit the much restored Gothic vaults at ground floor level which are used for exhibitions of contemporary Czech art. In the Grand Courtyard there is a modern statue of Hus by Karel Lidický.

The Carolinum is now the University Rectorate, the faculties being dispersed elsewhere around Prague.
Železná 9, Staré Město pedestrian zone. Metro to Můstek or Náměstí Republiky. Cloister and halls open: during exhibitions; otherwise visiting hours are 10am–6pm.

LETOHRÁDEK HVĚZDA (Star Castle)
The Imperial Governor of Bohemia, Ferdinand of Tyrol, himself designed this remarkable Renaissance country house. It was built in the form of a six-pointed star between 1555 and 1557 by Italian architects and has preserved its original aspect with the exception of the roof, which is 18th-century. An especially attractive feature inside are the stucco reliefs of scenes from classical mythology and Roman history. The gods personifying the planets named after them form the focal points on the ceilings of individual rooms. In the cellars are tableaux with explanatory texts about the Defenestration of Prague and the Battle of the White Mountain.

After World War II the castle was sensitively restored by Pavel Janák and now houses displays on the life and work of the historical novelist Alois Jirásek and the painter Mikoláš Aleš (see page 80).
Tel: 36 79 38. Open: Tuesday to Saturday 9am–4pm, Sunday 10am–5pm. Admission charge. Trams 8 and 22 to Vypich (beyond Břevnov).

Above: the aptly named Star Castle, now home to an artist's museum
Below: Exhibit at the Star Castle

Karlův Most

(Charles Bridge)

In high summer visitors throng the parapets and gangway of the Charles Bridge; neo-hippies take their ease; pretty girls sell knick-knacks; and often a Dixieland jazz band supplies free entertainment. In winter the freezing mist rises from the Vltava, turning the heavily wrapped figures hurrying across the bridge into shrouded spectres. Nowhere in the city is more atmospheric, or richer in historical associations.

Of all Charles IV's ambitious undertakings, which included the building of St Vitus and the founding of the university and Nové Město, this 516m-long sandstone bridge with its 16 graceful arches has most captured the imagination. He laid the foundation stone on 9 July 1357, but did not live to see the great work completed in 1383. That pleasure was reserved for his son Wenceslas IV (who murdered his Vicar-General, John of Nepomuk, by having him thrown from the bridge).

Over the years the Charles Bridge has witnessed many dramatic events. During the Middle Ages dishonest traders were suspended from the bridge in wicker baskets. In 1621 the heads of the executed Bohemian nobles, who had fought against the Habsburgs, were exhibited on the tower at the Staré Město end. The peace that put an end to

A new day dawns on the Charles Bridge

the Thirty Years' War in the Czech lands was signed in the middle of the bridge, the Swedish army having failed to take the Old Town.

ARCHITECTURE AND STATUARY

The Prague Bridge (it only became the 'Charles Bridge' in 1870) replaced an earlier Romanesque one made of stone and named after Judith of Thuringia, wife of Vladislav I. It is not quite straight, having been built using the Judith Bridge's land foundations on each side of the river, but with midstream piers placed slightly to the south of the previous construction.

St John Nepomuk, Prague's most revered saint

The bridge towers

All that is left from the Romanesque period are some piers sunk in the river-bed and the smaller of the two bridge towers at the Malá Strana end, the latter having been rebuilt in the Renaissance. It is joined to a higher tower built by King George of Poděbrady which is largely an imitation of the tower on the opposite bank.

At the Staré Město end is the impressive Gothic tower designed by Petr Parléř. St Vitus is portrayed on the east façade, flanked by Charles IV (on the left) and Wenceslas IV (on the right). Above them are St Adalbert and St Sigismund, patron saints of Bohemia, and below them the coats of arms of the Czech Crown Lands together with a veiled kingfisher, the heraldic symbol of Wenceslas IV.

The statues

A remarkable feature of the Charles Bridge is the baroque sculptures of saints along each side of it, erected between 1683 and 1714. Originally there had been only one stark crucifixion scene on the bridge, but in the late 17th century Bernini's sumptuous statuary for the Ponte dei Angeli in Rome inspired a similar plan for Prague.

The Jesuits were keen to promote the cult of St John Nepomuk and his statue by Jan Brokoff (in the middle of the north side) was the first to be put up in 1683. The Brokoff dynasty (Jan, Jan Michael and Ferdinand Maximilian) supplied a number of fine works.

Some of the older statues have been replaced by copies; several uninspired neo-Gothic works went up in the 19th century. The finest baroque sculpture is Matthias Braun's representation of St Luitgarde and the Crucifixion (1710) on the south side, the fourth from the Malá Strana bank. A modern work of some distinction is Karel Dvořák's *SS Cyril and Methodius* (1938) on the north side, the fifth from the Staré Město bank.
Trams 12, 17, 18 or 22. Metro to Staroměstská or Malostranská.

Stairway to heaven – St Salvator's façade is decorated with statues of angels and saints

KLEMENTINUM (Clementinum)
In 1556 Ferdinand I summoned to Prague 40 Jesuit monks. They took over the Dominicans' church of St Clement and founded the Clementinum, a centre of learning and propaganda for the faith that continued to expand up to the mid-18th century, swallowing up 32 houses, three churches, several gardens and even the heretical Carolinum (see page 66). After the disbanding of the Jesuits, the now non-heretical Charles University moved its library to the Clementinum, and that collection is today part of the Národní kníhovna (National Library) which boasts an estimated five million volumes, as well as numerous incunabula and manuscripts. Its most celebrated item is the *Codex Vyšegradensis* of 1085.

The architects chiefly involved in building the Clementinum were Carlo Lurago, Francesco Caratti and (later) František Kaňka.

Kostel sv Kliment (St Clement's Church)
This is the Prague base of the Uniate Church (halfway between Orthodoxy and Catholicism). Access is difficult, but it is worth persisting in order to see the frescos by Jan Hiebel depicting the life of St Clement, and the fine sandstone sculptures of the church fathers and the four evangelists by Matthias Braun. *Open: 5pm on Friday or apply at the sacristy.*

The Marian Chapel
František Kaňka completed this chapel in 1730. Its secular name of 'Hall of Mirrors' is a reference to the mirrors in the ceiling. It served as the private chapel of the Brotherhood of Our Lady, hence Hiebel's ceiling fresco depicting the life of the Virgin. Occasional concerts and exhibitions are held here from time to time.

Vlašská kaple (The Italian Chapel)
Built in 1597 for the community of Italian painters, sculptors and masons in Prague, the chapel has an elliptical form inspired by the Roman baroque. It still belongs to the Italian state.
Located between St Clements and the east end of the Church of the Holy Saviour on Karlova. Entry is difficult.

The halls
Kaňka and Hiebel designed and decorated the **Jesuit Library** or 'Baroque Hall' in the east wing, with its salomonic columns and ceiling fresco of *The Temple of Wisdom*. Also frescoed by Hiebel, the **Mathematical Hall** contains a collection of table clocks.

The **Mozart Room** is a full-blown example of rococo, notable for the high quality paintings and finely carved bookcases.
Access to the Klementinum is from Karlova or Křížovnická ulice. Trams 17 or 18 to Staroměstská. At the time of writing the halls and library are closed to the public for renovation. Group visits may be allowed. Tel: 2489 3111.

KŘÍŽOVNICKÉ NÁMĚSTÍ (Knights of the Cross Square)

The Coronation Route of the Bohemian kings passed along Karlova and across this, the Knights of the Cross Square, on to the Charles Bridge, before winding its way through the Lesser Quarter up to the Hrad. The square derives its name from the hospice Order of the Knights of the Cross with a Red Star, keepers of the Judith Bridge in the 13th century (see page 68).

Next to the bridgehead is Ernst Hähnel's cast-iron statue of Charles IV. The outbreak of the 1848 revolution in Prague prevented its planned inauguration on the 500th anniversary of the founding of Charles University in 1348.

Kostel sv Františka Serafinského (The Church of St Francis Seraphicus)
This was the knights' own church. For details see page 37. On the street corner is Jan Bendl's statue of St Wenceslas.

Plaque to Joseph II on the wall of the Klementinum

Kostel svatého Salvátora (The Church of the Holy Saviour)
This Jesuit church on the east side of the square forms part of the Clementinum (see opposite). It took over a century (1593–1714) to complete. The rich (now blackened and crumbling) statuary on the façade is by Jan Bendl. Carlo Lurago was the main architect, while the lavish stucco inside is the work of Domenico Galli. Karel Stádník's modern (1985) glass and metal altar symbolizing the cosmos harmonises surprisingly well with the interior.
Metro and trams 17 or 18 to Staroměstská.

The Loreta

After the Catholic victory at the battle of the White Mountain (1620), the Czech lands were swept by a wave of pietism. To eradicate nostalgia for Protestant heresies and heroes, the formidable propaganda machine of the Counter-Reformation was turned up full blast. In particular, the Marian cult was exploited, replacing Protestant hostility to images with a mixture of symbolism, sensuality and superstition.

Mariolatry was given its first major impetus in Bohemia with the founding of the Prague Loreto by the Spanish-born Benigna Kateřina Lobkowicz. The focus of the sanctuary was an imitation of the Santa Casa, claimed to be the historical house of the Virgin Mary, originally in Nazareth, and deposited by angels on Italian shores in a laurel grove (Loreto) near Ancona. This legend evidently appealed to the public imagination: countless 'Loretos' sprang up in Catholic Europe and eventually there were some 50 in Bohemia alone. The Prague Loreto was started in 1626, close to an existing Capuchin monastery.

The west façade

Designed by Christoph Dientzenhofer, the outer wing was completed by his son Kilián in 1726. The rich decoration on the façade is crowned by a kneeling Virgin Mary (surmounting the left-hand gable), complemented by the Angel of the Annunciation (over the right-hand gable). Below are the four evangelists and St Christopher. The coat of arms of the shrine's patrons, Prince Philipp Lobkowicz and his wife are above the entrance.

Baroque tower above the Loreta shrine

The clocktower

A carillon in the clocktower plays the Marian hymn 'We Greet Thee a Thousand Times' every hour on the hour. The mechanism was made in Amsterdam and consists of 27 bells collectively weighing some 1,600kg. The astronomer Tycho Brahe successfully petitioned Rudolf II to order the monks to ring their evensong carillon before it got dark, as otherwise it disturbed his concentration when stargazing!

The cloisters

The lower arcade has an upper storey added by Kilián Dientzenhofer in the 1740s. The painted ceiling vaults feature motifs from the Litany used during processions at Loreto itself. The arcades are lined with representations of wonder-working saints whose fields of expertise are inscribed below them: thus sore throats are the province of St Blaise, toothache of St Apollonia and plague of St Roch.

Kostel Narození Páně (Church of the Nativity)

On the east side of the sanctuary this 18th-century church has a fine fresco of Christ in the Temple by Václav Reiner. Other frescos by J V Schöpf depict the Christmas scene of the Three Kings and the Shepherds.

The Santa Casa

In the middle of the cloisters is the Santa Casa, the spiritual focus of the Loreto. It was built by Giovanni Orsi and Andrea Allio in 1631. Rich stucco inside depicts figures from the Old Testament and scenes from the Life of the Virgin. A lime-wood statue of Our Lady encased in elaborate silver decoration glimmers in the dim religious light.

The treasury

On the first floor is the fabulous treasury, whose loveliest work is a diamond monstrance (1698) designed by Johann Fischer von Erlach and made by Viennese silversmiths. It is a stunning example of baroque extravagance, a dramatic and sensual representation of the joint victory of Maria Immaculata and the Trinity over the forces of evil. It is said that many of its 6,222 diamonds came from the court dress of the benefactress, Countess Kolowrat, who left her entire fortune to Loreto.

Loretánské náměstí (Hradčany). Tram 22 to Památnik pisemnictvi. Tel: 2451 0789. Open: Tuesday to Sunday 9am–7pm. Admission charge.

Old Town Prague is a magnificent symphony of spires and towers

Malá Strana

*M*alá Strana, the 'Lesser Quarter', is at once intimate and grandiose. Crowded with sumptuous baroque palaces and churches, its narrow cobbled streets tail off into green garden-oases of silence, or snake towards the river through irregular medieval squares.

In 1257 the ambitious Přemysl Otakar II founded a 'New Town' in an area on the west bank of the Vltava hitherto consisting only of scattered communities. The latter included the Knights of St John, whose headquarters was beside the bridgehead of the Judith Bridge, as well as a few market traders and the Jews of Újezd. The knights were allowed to stay, while the Jews and the other inhabitants were summarily evicted to make way for German merchants. New towns, directly dependent on royal privilege, were vital for the king's exchequer and represented Otakar's attempt to outmanoeuvre the Czech nobility, the latter having won the right to set their own level of taxation.

Otakar's 'New Town' was given the name 'Lesser Quarter' under Charles IV, who founded his own 'Nové Město' on the opposite bank of the river. Thereafter the area suffered a decline, being extensively damaged in 1419 by the Hussites and further devastated by the great fire of 1541. From the ashes arose Renaissance houses, together with an Italian quarter established by the hundreds of Italian artisans attracted to Prague during the building boom under Ferdinand and Rudolf (names like Vlašská (Italian) Street recall their presence).

The Kinsky Palace – a bird's-eye view

Nerudova, where every door tells a story

THE WORLD OF JAN NERUDA

Nerudova, the street in Malá Strana named after the celebrated Czech writer and journalist, Jan Neruda (1834–91), rises sharply from Malostranské náměstí towards the Hrad. Near its summit, at no 47, is the house where Neruda was born and the streets round about are the settings for his *Tales of the Lesser Quarter* (1878). No other writer evokes the day-to-day life of 19th-century Prague with such a sure touch, casting an ironically sympathetic eye on petty bureaucrats, small traders, exploited drudges and impoverished students.The stories conjure a tragi-comic world replete with farcical misunderstandings and unfortunate accidents, with small tragedies and lonely deaths.

Each story is a sharp vignette of late-Habsburg Prague, with its bumbling censorship, divided loyalties and mixed cultures. Neruda, the son of a charwoman and a tobacconist, was in many ways typical of the milieu he describes. He fell foul of the conservative Czech nationalists, while his adherence to the 'Young Czech' cause made him suspect in the eyes of German officialdom. Deserted by friends and derided by his enemies he died unjustly discredited and neglected.

Malá Strana acquired its present aspect when Catholic nobles loyal to the Habsburgs were granted most of it in the 1620s, and built their fabulous palaces. The diminutive baroque town – it encompasses a mere 60 hectares – has survived the turmoil of recent history almost unscathed. The main scenes of Miloš Forman's film *Amadeus* were set in virtually the same town as that which was visited by Mozart in 1787, when he stayed with his patron Count Thun in what is now the British Embassy.

The Lesser Quarter's last great social upheaval was in 1948 when the Communists solved the housing shortage by dividing its palaces into apartments. Thereafter the world's potentially most glamorous council houses slowly decayed for 40 years. The restitution law of 1990 will restore some buildings to their original owners, while no doubt others will be taken over by private concerns that have the funds to restore them.

Malá Strana. Trams 12, 18 and 22 to Malostranské náměstí; metro to Malostranská.

IMAGES OF PRAGUE

Eloquent images of Prague have been captured by two great Czech photographers – Josef Sudek (1896–1976) and František Drtikol (1878-1961). Sudek is known for his brilliant series of shots (1924) documenting the final phase in seven centuries of building St Vitus's cathedral. Drtikol's cityscapes range from winter panoramas of ice-clad bridges to the covered stairways and secret alleys of the city.

Even more evocative are the haunting scenes painted by Jakob Schikaneder (1855–1924). *Nocturne in Prague* (1911) is typical: on a chill autumn evening a woman and child hurry into the shadows out of the glare of a gas-light. Schikaneder's is a world of eventide and shadowy figures, of dark buildings splashed with light, frozen cabmen waiting in snow-covered squares.

Powerful in a different way are the three expressionistic Prague landscapes by Oskar Kokoschka (1886–1980). The turbulent brush-strokes and the bold angle of vision are most striking in his version of *Charles Bridge and the Hradčany* (1935).

In the work of Karel Chaba (born 1925), naïve and surrealist elements fuse in a series of kaleidoscopic dreams: the city, drenched in colour, seems to dance a crazy Czech polka along the banks of the Vltava. The

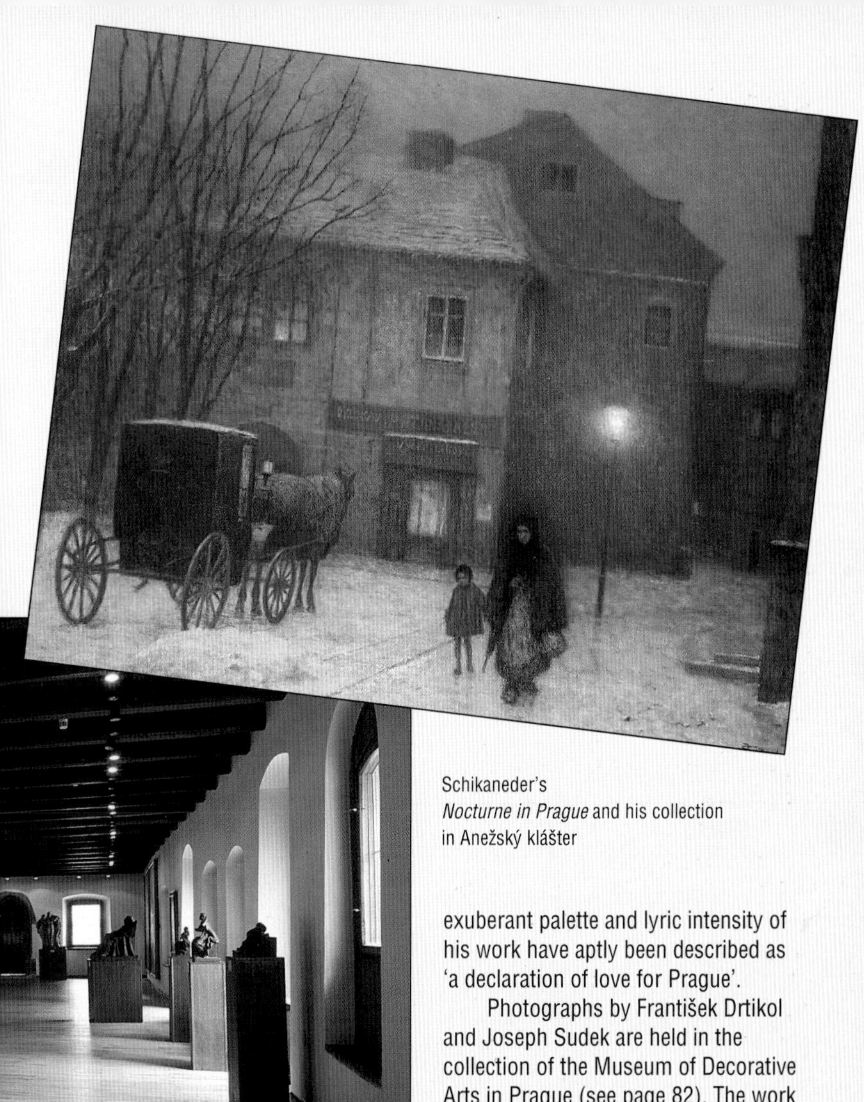

Schikaneder's
Nocturne in Prague and his collection
in Anežský klášter

exuberant palette and lyric intensity of his work have aptly been described as 'a declaration of love for Prague'.

Photographs by František Drtikol and Joseph Sudek are held in the collection of the Museum of Decorative Arts in Prague (see page 82). The work of Jakob Schikaneder is in the Anežský Klášter part of the National Gallery Collection (see pages 90–1); Oskar Kokoschka's landscapes are in the Šternberský palác part of the National Gallery. Galerie Victoria Art (Mostecká 6, open: 10am–10pm) handles the work of Karel Chaba.

Malostranské Náměstí

Once the outer bailey of Prague castle, the Lesser Quarter Square became the focus of the small town founded by Otakar II in 1257. In the Middle Ages a Romanesque rotunda dedicated to St Wenceslas stood in the middle of it, along with a pillory and a gallows. The Town Hall was established in the 15th century and later a parish church of St Nicholas (Kostel sv Mikuláše) was built, dividing the upper and lower halves of the square.

The Lower Square

Gothic, Renaissance and baroque elements mingle gracefully in the lower part of the square. To the west it is closed by the rear walls of St Nicholas and the former Jesuit College. In front of these is the rococo Grömling Palace (built by Josef Jäger in 1773); the Radetzky Café (named after the Austrian general who put down the Italian rebellion against the Habsburgs in 1848) was opened on the ground floor, and became a haunt of the Prague literati, including Kafka, Brod and Werfel. The name 'Radetzky' did not survive the rise of Czech nationalism – the café is now the Malostranská kavárna.

Watch the world go by outside the Town Hall

On the east side of the square, at no 21, is the former Town Hall (Malostranská radnice). It was here that Protestant groups hammered out the Bohemian Confession of 1575, which guaranteed freedom of religion (although the Habsburgs subsequently broke their undertakings).

On the north side of the Lower Square are the Sternberg Palace (no 19), where the great fire of 1541 broke out, and the Smiřický Palác (no 18), where the dissident Czech nobles plotted the overthrow of Ferdinand II's hated Catholic councillors on 22 May 1618. On the following day they matched words to actions by throwing the councillors out of the windows of the Bohemian Chancellery (the third Defenestration of Prague).

The Upper Square

At the centre of the square is Giovanni Alliprandi's Plague Column (1715). The whole of the west side is occupied by the Liechtenstein Palace, whose neo-classical façade dates to 1791. The Liechtensteins lost their Czech possessions (about 10 times the size of the present Grand Duchy) in 1918; unfortunately for them the recent restitution law covers only what had been grabbed by the state after 1948.

Kostel Sv Mikuláše (Basilica of St Nicholas)

This beautiful basilica, opposite the Liechtenstein Palace, is regarded by many as the finest church in Prague, if not in Central Europe. It was commissioned by the Jesuits from Christoph Dientzenhofer, who was responsible for the elegant façade (1710). Christoph also completed the nave, the side-chapels and the galleries before his death in 1722, while his son Kilián Ignác built the choir and the ambitious dome (1752); the townspeople refused to enter the church until a commission of experts had pronounced it safe.

An outstanding feature of the pink, pastel green and cream-coloured interior is Lukáš Kracker's enormous (1,500sq m) illusionist fresco on the ceiling of the nave, showing the Apotheosis of St Nicholas. The outsized statues of the early fathers of the church under the cupola are by Ignaz Platzer, who also sculpted the St Nicholas above the high altar. The superb rococo pulpit (1765) by Richard and Peter Práchner is decorated with allegories of Faith, Hope and Charity; rather oddly juxtaposed to these is a scene of John the Baptist being beheaded. The drama, richness and scale of St Nicholas make a visit here one of the most moving experiences for any visitor to Prague.

Open: 9am–5pm. Admission charge. Trams 12 and 22 to Malostranské náměstí. Metro station: Malostranská.

High baroque sculpture inside the Basilica of St Nicholas

Museums

*T*he restoration of democracy in Czechoslovakia has resulted in visitors being denied a number of surreal museum experiences that were formerly available. Where are they now, the V I Lenin Museum and the Klement Gottwald Museum, where bemused western visitors were once subjected to bombastic harangues? Answer: in the dustbin of history, although there are apparently plans to reconstitute their collections in a new (and doubtless coldly analytical) 'Museum of Communism'. Other idiosyncratic displays have been sanitised.

The following list covers those of Prague's remaining minor museums that are likely to be of interest to the casual visitor. Two that were closed at the time of writing (either for renovation or for possible relocation) are featured, in the hope that they will be functioning again before too long.

MUZEUM ALOISE JIRÁSKÁ A MIKOLÁŠE ALEŠ (Alois Jirásek and Mikoláš Aleš Museum)
The novelist Jirásek (1851–1930) and the painter Aleš (1852–1913) were two luminaries of the 19th-century Czech national revival. The most interesting part of the show are the book illustrations by Aleš on the first floor (see page 67).
Letohrádek Hvězda, Liboc. Tel: 36 79 38. Open: Tuesday to Sunday 10am–5pm. Admission charge. Trams 8 and 22 to Vypich.

FRANZ KAFKA PERMANENT EXHIBITION
Kafka's life and works are shown in words and pictures. This is a first step in the ambitious programme of the Kafka Centre: a museum, a congress centre, an archive, a publishing house, a theatre, a club, a bookshop and a coffee-house are just some of the other projects planned. Clearly we are witnessing the birth of a Kafka industry.
U Radnice 5, just off Staroměstské náměstí. Open: daily 10am–6pm. Tram and metro to Staroměstská. The booklet Franz Kafka a Praha *(with English text) is obtainable here and details all the places in Prague associated with the writer.*

MUZEUM LETECTVÍ A KOSMONAUTIKY (Museum of Flying and Cosmonauts)
Some 80 of the 200 aeroplanes in this museum are on display. There is also information about Czechoslovakia's participation in space research. The capsule in which the first Czech cosmonaut landed (1978) may be seen, and the cosmonaut himself (V Remek) is the current director of the museum.
Kbely (Prague 9). Tel: 82 47 09. Open: 1 May to 30 October, Tuesday to Sunday 10am–6pm. Admission charge. Metro: line B to Českomoravská, then buses 185 and 259 to Leteckví muzeum.

LOBKOWICZ PALÁC NARODNÍ HISTORICKÉ MUZEUM (Lobkowicz Palace National History Museum)
A visit to this museum is best included in a tour of Castle Hill. Exhibits include copies of the Bohemian Crown Jewels,

and St Wenceslas's sword, together with material relating to the Hussite wars and the Battle of the White Mountain.
Jiřská 3, Hradčany. Tel: 53 73 06. Open: Tuesday to Sunday 9am–5pm. Admission charge. Tram 22 to Pražský hrad; metro to Malostranská.

NÁPRSTKOVO MUZEUM (Náprstek Museum)
The Náprstek Museum is closed for reorganisation at the time of writing. It contains artefacts from American Indian, Oceanian, African and Asian cultures.
Betlémské náměstí 1, Praha 1. Tel: 2421 4537. Open: Tuesday to Sunday 9am–noon, 12.45–5pm. Admission charge. Trams 6, 9, 18 and 22 to Národní divadlo or metro to Národní třída.

NÁRODNÍ TECHNICKÉ MUZEUM (National Technical Museum)
In a huge main hall this museum offers a marvellous display of historic transport, from Imperial railway carriages to early sports cars and aeroplanes. Space is also devoted to chronography, photography, astronomy, mining and much else.
Kostelní 42, Letná. Tel: 37 36 51. Open: Tuesday to Sunday 9am–5pm. Admission charge. Trams 1, 8, 25 and 26 to Letenské náměstí.

NÁRODOPÍSNÉ MUZEUM (Ethnographic Museum)
This substantial collection of artefacts and folk costumes, including models of peasant houses, is currently closed for renovation and possible relocation.
Petřínské sady 98 (Kinsky villa). Try telephoning 53 11 51 for information or enquire at the National Museum (tel: 2423 0485). Trams 6, 9 and 12 to Náměstí Kinských.

Transport is the main theme of the National Technical Museum

PAMÁTNÍK NÁRODNÍHO PÍSEMNICTVÍ (National Literature Museum)
Housed in the Strahov Monastery, the collection covers Czech literature from Cyrillic bibles to the patriotic revival of the 19th century, together with works by writers in exile. The recently returned Premonstratensian monks are restricting the exhibition to temporary displays of selected material while they try to reconcile the needs of the contemplative life with a daily flood of visitors.
Strahovské nádvoří 1. Tel: 2451 1137. Open: Tuesday to Sunday 9am–12.30pm, 1–5pm. Admission charge. Tram 22 to Památník písemnictví.

PEDAGOGICKÉ MUZEUM JANA AMOSE KOMENSKÉHO (Jan Comenius Pedagogical Museum)

This display on the life and work of the humanist scholar Jan Comenius (1592–1670), known as 'the teacher of the nation', is the most interesting part of an otherwise rather turgid offering. There is also a memorial here to the philosopher Jan Patočka, who died in 1977 while being interrogated about his membership of Charter 77.

Valdštejnské náměstí 4. Tel: 53 62 74. Open: Tuesday to Sunday 9am–noon, 1–5 pm. Admission charge. Trams 12 and 22 and metro to Malostranská.

MUZEUM POŠTOVNÍ ZNÁMKY (Postage Stamp Museum)

This rather charming and little frequented museum has a stamp collection on the ground floor and some homely prints upstairs demonstrating that a postman's lot is not necessarily a happy one. The showrooms' frescos by Josef Navrátil – landscapes and genre pictures, together with themes from Bohemian legend – are a bonus.

Várů dům, Nové mlýny. Tel: 231 20 60. Open: Tuesday to Sunday 9am–5 pm. Admission charge. Trams 5, 14 and 26 to Revolucní.

MUZEUM TĚLESNÉ VÝCHOVY A SPORTU (Museum of Sport and Physical Culture)

The exhibition celebrates the achievements of great Czech sportsmen such as the legendary Emil Zátopek who set six new world records in the 11 years up to 1952 and won four Olympic gold medals. The other main feature is the history of the Sokol movement, founded by Czech nationalists in 1862 on the model of the German *Turnverband*. The idea of mass gymnastic displays was taken over by the Communists for their five-yearly 'Spartakiáda'.

Újezd 40. Tel: 2491 0172. Open: Tuesday to Sunday 9am–5pm. Admission charge. Trams 12 and 22 to Hellichova.

UMĚLECKOPRŮMYSLOVÉ MUZEUM (Museum of Decorative Arts)

Only a tiny proportion of the museum's vast horde of Renaissance to 19th-century treasures is on permanent display. A great deal of that is furniture, including some fine escritoires, cabinets and clocks. If there is a temporary exhibition drawn from the museum's particularly rich archive of posters and photographs, be sure not to miss it (see pages 76–7). The glass collection on the second floor (see pages 150–1) should hopefully soon be open again after reorganisation.

Ulice 17, listopadu 2. Tel: 2481 1241. Open: Tuesday to Sunday 10am–6pm. Admission charge. Trams 17 and 18 and metro to Staroměstská.

VOJENSKÉ MUZEUM (War Museum)

This military history museum is an engrossing collection located in the Schwarzenberg Palace. The period covered is from the Middle Ages to 1918 and a great deal of Bohemian history can be gleaned from the well-documented objects on display.

Hradčanské náměstí 2. Tel: 53 64 88. Open: Tuesday to Sunday 8.30am–5pm.

Admission charge. Tram 22 to Pražský hrad.

VOJENSKÉ MUZEUM
(War Museum)

Formerly the Museum of the Resistance and the Czechoslovak Army, this has now been revamped to make it more informative than its predecessor. Topics covered include the story of the Czech Legions in World War I and the assassination of the Nazi 'Reichsprotektor', Richard Heydrich, during World War II (see pages 37 and 142).
U Památníku 2, Žižkov. Tel: 27 29 65. Open: April to October, Tuesday to Sunday 10am–6pm; November to March, Monday to Friday 8.30am–5pm. Admission charge. Buses 133 and 207 to Pod památníkem. Metro to Florenc.

The Museum of Decorative Arts

The neo-Renaissance National Museum, built between 1885 and 1890 by Josef Schulz

MUZEUM HLAVNÍHO MĚSTA PRAHY (Museum of the City of Prague)

This museum contains substantial archaeological finds from earliest times, plus impressive statuary and sculpture, ranging from a Gothic Madonna of 1383 to a fine bronze of Hercules by Adrian de Vries. The baroque wood-sculptures are by the finest Bohemian masters, including Jan Bendl, Ferdinand Brokoff and Matthias Braun.

A curiosity is Antonín Langweil's model of historic Prague, which took him eight years to make between 1826 and 1834. Some 2,228 buildings have been lovingly recreated on a scale of 1:423, giving a unique picture of the city in the early 19th century.

Na Poříčí 52. Tel: 2422 3180. Open: Tuesday to Sunday 10am–12.30pm, 1.30–6pm. Admission charge. Metro and trams 3, 8 and 24 to Florenc.

NÁRODNÍ MUZEUM (National Museum)

The original project for the National Museum managed to unite the aspirations of the Czech and the German populations, but by the time the present heavy neo-Renaissance building (the Museum's third home) was completed in 1891, it had become a pugnaciously Czech affair. Inside are somewhat forbidding departments of zoology, botany, archaeology, palaeontolgy and mineralogy to wade through. Not for the casual visitor

Like a phoenix from the ashes, Prague's sumptuous National Theatre

The Pantheon

The focus of the building is the heroic marble Pantheon, a huge, richly decorated space under the cupola with statues and busts of famous Czechs and lunettes painted by František Ženíšek and Václav Brožík showing events from the nation's past. Six over-lifesize statues are placed next to the pillars: four giants of Bohemia's past – Jan Hus, Jan Comenius, František Palacký and Tomáš Masaryk; plus two comparative lightweights – Count Šternberk, the founder of the museum, and the writer Jan Neruda.

Václavské náměstí 68. Tel: 2423 0485. Open: Wednesday to Monday 9am–5pm. Admission charge. Metro to Muzeum.

NÁRODNÍ DIVADLO (National Theatre)

František Palacký was the moving spirit behind the founding of the National Theatre, Prague until then having had only German theatres that sometimes staged Czech plays. Josef Zítek's building was destroyed by fire even before it officially opened; it was rebuilt between 1881 and 1883, preserving his neo-Renaissance design by his pupil, Josef Schulz.

It is lavishly decorated inside with allegorical and historical themes by leading artists of the time. The New Scene, a rear extension built in 1983, was described ecstatically by one writer as a 'translucent honeycomb', although 'a sugar cube coated in phlegm' might be nearer the mark.

Národní třida 2. Tel: 2491 3437. Performances usually begin at 7pm. Trams 6, 9, 17, 18 and 22 to Národní divadlo.

TOMÁŠ GARRIGUE MASARYK

In its short existence (1918–92) Czechoslovakia produced two statesmen who won the respect of the world: Václav Havel and Tomáš Masaryk (1850–1937). Masaryk was of humble origin – an illegitimate child brought up by Slovakian peasants. He rose to become a professor of philosophy and a Social Democratic MP in the Reichsrat (Parliament) of the Austro-Hungarian Empire. In exile during the World War I he laid the foundations for the new Czechoslovak Republic, which proved to be a model of stability and democracy under his presidency (1918–35). In today's chauvinistic climate it is appropriate to recall that Masaryk stood for dignified love of one's country and abhorred the rabble-rousing nationalism of demagogues.

The Magic Lantern is an experience unique to Prague

'What is life? An illusion, a shadow, a story ...' wrote the 17th-century Spanish dramatist Calderón, a reflection which might well be applied to Prague's *avant-garde* illusionist theatres.

The cunning use of spectacle, illusion and music has its roots at least as far back as the propagandist art and drama sponsored by the Jesuits in the baroque age. Modern technology opened the way to a new form of multi-media spectacle first developed by Alfréd Redok in the 1950s. His Laterna Magika (Magic Lantern) shows made sophisticated use of lighting and film projection, and won world acclaim at the Brussels Expo of 1958. The idea has proved enduringly successful at the box-office. Its more or less imitative spin-offs include Theatrum Novum, Laterna Animata and the brilliant Ta Fantastika Black Light Theatre of Prague.

The show is a mix of live acting (mime, ballet) and scenic *coups de théâtre* brought about by film or slide projection and dramatic lighting effects. The third ingredient is music. Particularly successful are reworked myths, such as that of the Odyssey in terms of 'the common dawn and early childhood of today's Europe'.

It has to be admitted that there are times when gimmickry overwhelms the drama. On the other hand, the

LATERNA MAGIKA

fashionable sneer that Laterna Magika is merely a faked theatrical 'happening' put on for visitors misses the point entirely. The shows are popular because they are exciting, brilliantly staged and unique to Prague.

Productions recently in the repertoire include *Magic Circus*, *The Minotaur*, *Odysseus* and *Carmina Burana*. *Odysseus* is so overbooked that it is now staged in the Palace of Culture in Pankrác; the other shows usually take place in the Nová Scéna of the National Theatre.

For booking arrangements see **Theatre** on page 152.

Music Museums

Statue of Bedřich Smetana

BERTRAMKA

This charming 17th-century villa which houses the Mozart Museum is named after its second owner, František Bertram. However, it owes its fame to the fact that Mozart stayed here with his friends Josefa Dušková, the opera singer, and her composer husband, František. Mozart was allegedly locked into one of the rooms until he had completed a long-promised aria for Josefa and may also have completed the overture to *Don Giovanni* here.

There is an exhibition of the composer's life and work in the villa and a bust of him by Tomáš Seidan in the garden. Charming concerts of Mozart arias are held in the villa.

Mozartova 169, Smíchov. Tel: 54 38 93. Open: daily 9.30am–6pm. Admission charge. Trams 4, 6, 7 and 9 to Bertramka or metro (line B) to Anděl and 1.5km walk.

MUZEUM ANTONÍNA DVOŘÁKA (Antonín Dvořák Museum – Vila America)

The name of this gem of a baroque villa, built by Kilián Ignác Dientzenhofer, and with statues from Matthias Braun's workshop in the garden, refers to a hotel that used to stand near by, not to Dvořák's famous *New World Symphony*. The Dvořák collection includes musical scores and correspondence. One of the greatest of the 19th-century romantic composers, Antonín Dvořák (1841–1904) derived much inspiration from Czech folk music.

Ke Karlovu 20. Tel: 29 82 14. Open: Tuesday to Sunday 10am–5pm. Admission charge. Trams 4, 6, 16 and 22; metro to I P Pavlova.

MUZEUM BEDŘICHA SMETANY (Bedřich Smetana Museum)

A modern statue of the great Czech composer, sitting under a willow with his back to his beloved Vltava, is almost the only item of note in this otherwise undistinguished museum. Smetana (1824–84) composed the music most closely identified with the patriotic aspirations of Bohemia. His best known works are *The Bartered Bride* and *Dalibor* and his tone poem *My Country* contains a famous passage beautifully evoking the surging waters of the Vltava.

Novotného lávka 1. Tel: 2422 9075. Open: Wednesday to Monday 10am–6pm. Admission charge. Trams 17 and 18 to Karlovy lázně; metro to Staroměstská.

MUZEUM HUDEBNÍCH NÁSTROJŮ (Musical Instruments Museum)

Hitherto in the Grand Prior's palace in the Lesser Quarter, the museum is currently closed pending relocation to a more secure building after Stradivarius violins were stolen from it in 1990.

For information tel: 2451 0114/5.

MOZART IN PRAGUE

Mozart arrived for his second visit to Prague in the autumn of 1787 with a new opera under his arm. *The Seraglio* and *The Marriage of Figaro* had already brought him fame with the musically sophisticated Prague public; *Don Giovanni*, premièred on 29 October at the Estates Theatre (see page 117), was an even greater triumph. Legend has it that apprentices and servants whistled the arias in the streets. This was the high point of his success in Prague. *La Clemenza di Tito*, premièred in the autumn of 1791, was considered old-fashioned and was not well received (one aristocratic lady dismissed it as 'German hogwash'). Back in Vienna, Mozart died two months later on 5 December, virtually unmourned by his compatriots. However, Prague staged a grand memorial service for him on 14 December in the Kostel sv Mikuláše (St Nicholas Church) of the Lesser Quarter. The church, with a capacity for 4,000 people, was filled to overflowing and carriages blocked the surrounding streets. Josefa Dušková led the requiem singers and, in the words of a newspaper report: 'A thousand tears flowed in grieving memory of the artist, who had so often filled every heart with vital feeling through the matchless beauty of his music'.

The Dvořák Museum, housed in a baroque villa designed by Kilián Dientzenhofer

Stálé Expozice Národní Galérie

(National Gallery Collections)

In the late 18th century wealthy Prague burghers joined with Bohemian nobles to found the Society of Patriotic Friends of the Arts (1796), which included in its activities the collecting of painting and sculpture. From this modest beginning the National Gallery of Bohemia was to grow to its present impressive collections, now dispersed around the city.

ANEŽSKÝ KLÁŠTER (St Agnes Convent) Czech Art of the 19th Century

The artists of the *národní obrození* (national revival) are little known outside their native land. Antonín Mánes painted romantic landscapes or topographies redolent of an heroic past (for example Kokořín Castle). His more gifted son, Josef, had a much wider range, painting portraits and academic pictures as well as allegorical subjects. Mikoláš Aleš (see page 80) is represented with a number of works, including his cartoons for the *Legend of My Country* created for the National Theatre. Josef Navrátil's still-lifes are a high point of the exhibition, as are the superb evocations of Prague by Jakob Schikaneder (see pages 76–7).

U milosrdných 17. Tel: 2481 0628. Open: Tuesday to Sunday 10am–6pm. Admission charge. Metro to Staroměstská or Náměstí republiky, then trams 5, 14 or 26 to Révoluční.

Medieval sculpture in the National Gallery

KLÁŠTER SV JIŘÍ NA PRAŽSKÉM HRADĚ (St George's Convent at Prague Castle) Collection of old Bohemian art

The gallery is divided into two sections: Gothic art in the basement and at ground-floor level, baroque works on the first floor. Among the former are fine examples of International Gothic or 'Beautiful Style'. One of the earliest free-standing sculptures (1373) is a bronze of *St George and the Dragon.*

High points are the nine-panelled altarpiece from the Cistercian monastery at Vyssí Brod and paintings of saints by Master Theodoric. Arresting baroque works include Bartolomaeus Spranger's *The Risen Christ* and sculptures by Maximilian Brokoff and Matthias Braun, in particular the latter's sculpture of *St Jude.*

Jiřské náměstí 33. Hradčany. Tel: 2451 0695. Open: Tuesday to Sunday 10am–6pm. Admission charge. Tram 22 to Pražský hrad; metro to Malostranská.

ŠTERNBERSKÝ PALÁC (Sternberg Palace)
Old European art, French 19th- and 20th-century art
Albrecht Dürer's *The Feast of the Rose Garlands* is the gallery's most celebrated possession, but there are fine works also by Brueghel, El Greco, Rembrandt, Rubens, Caspar David Friedrich, Gustav Klimt, Egon Schiele and Edvard Munch to name but a few. The separate collection of French 19th- and 20th-century painting is of extremely high quality and contains works by Cézanne, Pissarro, Gauguin, Manet, Degas, Seurat, Signac, Derain, Braque and Picasso (see page 59).
Hradčanske náměstí 15. Tel: 2451 0594. Open: Tuesday to Sunday 10am–6pm. Admission charge. Tram 22 to Pražský hrad; metro to Malostranská.

ZÁMEK ZBRASLAV (Zbraslav Castle)
Collection of 19th- and 20th-century Czech sculpture
It is well worth the trek to this former monastery where works by Josef Myslbek, Otto Gutfreund and Stanislav Sucharda are on show. Jan Štursa's nudes (*Eve* and *Melancholy Girl*) are particularly impressive. In 1994 it is planned to bring the collection of Asian Art to the castle.
Zbraslav nad Vltavou, Prague 5. Tel: 59 11 88 – 9. Open: Tuesday to Sunday 10am–6pm. Admission charge. Metro (line B) to Smíchovské nádraží; then buses 129, 241, 243, or 255 to Zbraslavské náměstí.

There are plans to install the National Gallery's Collection of 20th-century art in the Veletržní Palác before the end of 1993. Ring one of the other branches to check the position.

Zbraslav Castle, now the Museum of Sculpture. The building was converted from a monastery, which in turn had been Otakar II's hunting lodge

MUSICAL PRAGUE

'Music,' according to an old Prague saying, 'was born of necessity and became a necessity.' In earlier times musicianship offered one of the few possibilities of climbing up the social ladder. The prize for success was noble patronage (very attractive), and (even more attractive) escape from serfdom or 14 grim years of compulsory military service. By the 18th century Prague musicians were famous all over Europe.

Prague audiences are traditionally more musically literate than elsewhere. As early as the 1720s the public could buy cheap seats in private theatres, where censorship was also lax compared with the stiff court opera in Vienna. Thus the people of Prague understood and loved the music of Mozart at a time when Vienna turned its back on him.

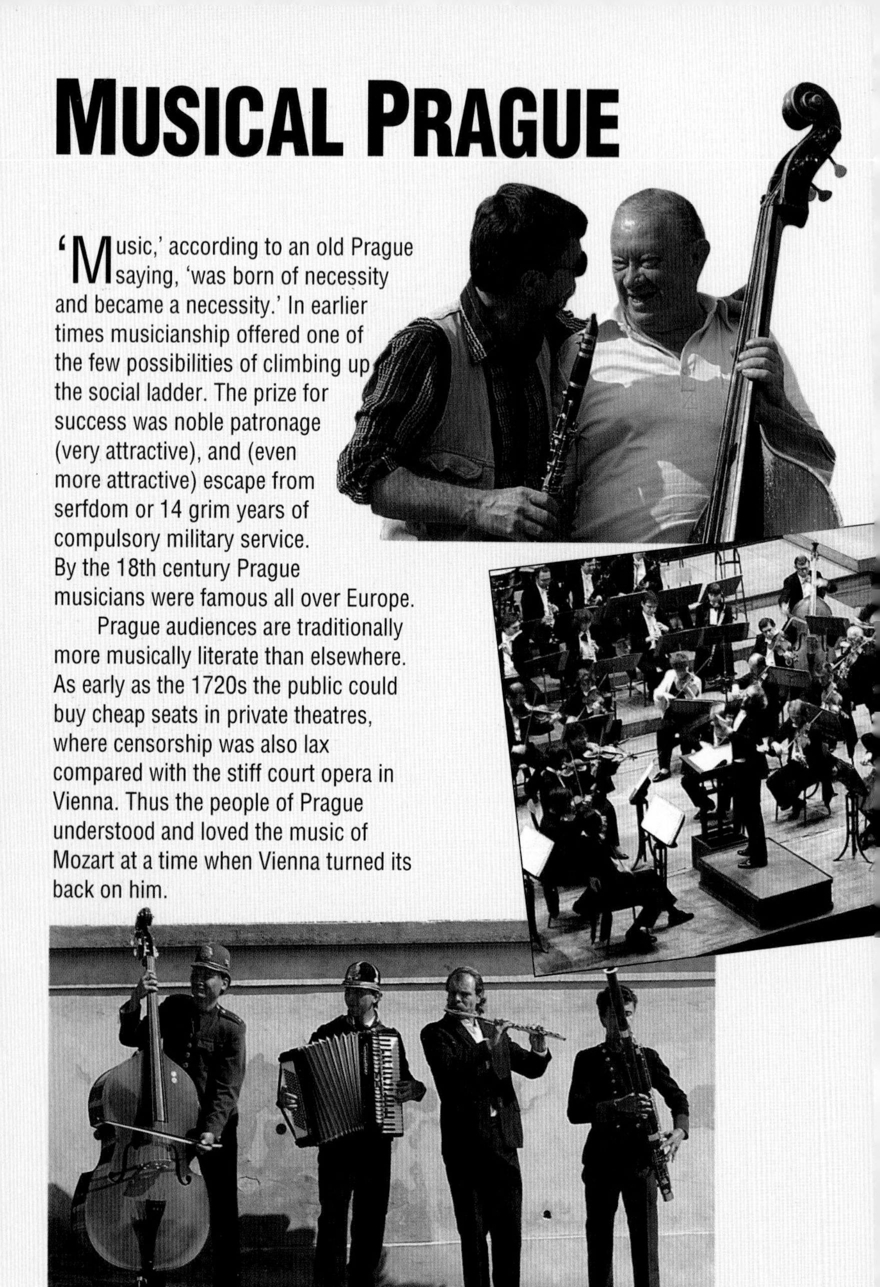

From classical to pop, Prague caters for all musical tastes

Nowadays Prague's spring and autumn music festivals attract music-lovers from all over the world. Throughout the year churches, former convents and baroque palaces ring to the sound of sacred and profane harmonies. The giants of Czech music, Bedřich Smetana (1824–84), Antonín Dvořák (1841–1904) and the Moravian Leoš Janáček (1854–1938) tend to dominate the symphony programmes. Opera and symphony concerts are held in the great auditoria built in the late 19th-century boom of patriotic culture: the National Theatre (Národní divadlo – 1881), the Rudolfinum (1884) and the Community House (Obecní dům, 1911); but many delightful works by minor composers of the Czech baroque are played in such romantic settings as Bazilika svatého Jiří (St George's Basilica) on Castle Hill.

In the 18th century the English musicologist, Charles Burney, gave Prague the title of 'The Music Conservatory of Europe'. Were he to come back now, he would see no reason to change his opinion. Hurrying figures with cello cases are still to be seen on the streets, and on languid summer afternoons melodious strains waft from many an open window.

Palaces

The majority of Prague's noble palaces are today occupied by government institutions, museums or embassies. Most are the product of the building boom in baroque times.

After the Battle of the White Mountain in 1620, German aristocrats and war profiteers loyal to the Habsburgs took over property that had belonged to the now exiled Bohemian nobility, or bought up land at depressed prices. The most spectacular example of megalomaniac building from this period is the huge Valdštejnský palác (Wallenstein Palace) in Malá Strana (see page 96); to make way for it 23 houses, three gardens and a brickworks were demolished.

A baroque palace was designed to display its owner's wealth and importance – the secular equivalent of baroque church architecture with its fabulous ornamentation. This period of ostentatious noble display lasted until the second half of the 18th century, when centralising reforms under Maria Theresa and Joseph II reduced the power and importance of feudal lords. Few palaces were built after about 1750, although a number of existing ones were rebuilt or refurbished to reflect contemporary taste.

The palaces listed below have been chosen for their aesthetic or historical interest. The list excludes those described in the contexts of Hradčanské náměstí (see pages 58–9), Malostranské náměstí (see pages 78–9) and Staroměstské náměstí (see pages 108–11).

CLAM-GALLASOVSKÝ PALÁC (Clam-Gallas Palace)

The palace was built between 1713 and 1719 by Giovanni Canevalle to a design by the great Viennese architect Johann Bernhard Fischer von Erlach. For such a noble building it may seem rather cramped in its surroundings on Husova, and indeed the owner had assumed that he would be able to demolish the block opposite so that his palace would look on to a square. Not surprisingly the residents had other ideas, so that a mortified Count Gallas had to be content with what he had.

Notable are the Atlas figures of the two doorways, the work of Matthias Braun, who also made the rest of the sculptural decoration. Inside there is a grand staircase with stucco by Santo Rossi and above it a fresco (*The Triumph of Apollo*) by Carlo Carlone. The city archives are now housed in the palace, which means that you can usually wander in to have a look.

Husova ulice 20, Staré Město. Trams 17 and 18 to Staroměstská. The palace is open for researchers in the city archives.

ČERNÍNSKÝ PALÁC (Černin Palace)

Four generations of Cernins and as many architects worked on this enormous palace, (its façade is 135m wide). Jan Černin who conceived the project, was ambassador in Venice, where he is said to have persuaded Bernini to do the first sketch for the building. It was not completed until 1720. Since 1932 it has belonged to the Ministry of Foreign Affairs. In 1948 Jan Masaryk, the only non-Communist in the government, mysteriously 'fell' to his death from its upper floor.

Loretánské náměsti 5, Hradčany. Tram 22 to Památník písemnictví. The palace is not open to the public.

LOBKOWICKÝ PALÁC (Lobkowicz Palace)

The powerful Lobkowicz family ended up with no less than three palaces in Prague, of which this is the most aesthetically pleasing, a masterwork by Giovanni Alliprandi (1707) with a further storey added in 1769 by Ignaz Palliardi.

The best view of its baroque splendour is from the far side of the English landscape garden at the rear (the palace is now the German Embassy, but access to the garden is sometimes possible). David Černý's bizarre sculpture of a gold-painted Trabant here, bearing the title 'Quo Vadis?', commemorates the time in 1989 when hundreds of East Germans occupied the embassy grounds, demanding West German citizenship.

Vlašská ulice 19, Malá Strana. Trams 12 and 22 to Malostranské náměstí. Open to those on embassy business.

The Černín Palace now houses the Czech Foreign Ministry

MORZINSKÝ PALÁC (Morzin Palace)

A striking feature of the Morzin Palace's façade are the two muscle-bound figures of Moors supporting the balcony. This is a reference to the family name, Morzin – 'Moor' in Czech. The architect Giovanni Santini-Aichel created the palace out of three older houses in 1714. All the sculptural decoration is by Ferdinand Brohoff: above the two side doorways are allegories of *Day* and *Night*, while the *Four Corners of the World* are represented on the roof. The building is now occupied by the Romanian Embassy.
Nerudova ulice 5, Malá Strana. Trams 12 and 22 to Malostranské náměstí. Open to those on embassy business.

NOSTICKŮV PALÁC (Nostic Palace)

The Nostic family, great patrons and collectors of art in the 17th century, commissioned Francesco Caratti to build this sumptuous palace in 1658. Additions and alterations were made by Giovanni Santini-Aichel in the 18th century: he built the top storey, to which Michael Brohoff added statues of Roman emperors. The Nostic family built up their own picture gallery, and a library which is still in existence. The building is now shared between the Dutch Embassy and the Ministry of Education. Scenes from the film *Amadeus* were shot in the palace's baroque interiors.
Maltézské náměstí 1, Malá Strana. Trams 12 and 22 to Hellichova. Scholars may visit the library.

THUN-HOHENŠTEJNSKÝ PALÁC (Thun-Hohenstein Palace)

This vast palace is a good example of accretive building in Malá Strana, whereby rich owners of small plots of land relentlessly extended their residences by buying up their less wealthy neighbours. The present building dates to 1726 and was designed by Giovanni Santini-Aichel. The eagles with outstretched wings over the entrance (by Matthias Braun) are the heraldic birds of the Kolowrat family, who originally built the palace (the Thun-Hohensteins inherited it only in 1768). The Italian Embassy now occupies the building.
Nerudova ulice 20, Malá Strana. Trams 12 and 22 to Malostranské náměstí. Open to those on embassy business.

VALDŠTEJNSKÝ PALÁC (Wallenstein Palace)

Albrecht von Waldstein (in English 'Wallenstein') was one of the great opportunists of history. Although he came of Protestant stock, he served the Habsburgs in the Thirty Years' War, rising swiftly to become Commander of the Imperial Armies. Having amassed a gigantic fortune he decided to build the grandest palace in Prague, and

Baroque features of the Wallenstein Palace

Tranquillity reigns in the attractive gardens of the Wallenstein Palace

demolished a large slice of the Lesser Ouarter to do so. The resulting palace remained in the hands of the Wallenstein family until 1945.

The Wallenstein Palace was built between 1624 and 1630 to plans by Andrea Spezza. It encompassed five courtyards and a spectacular garden and was surrounded by a high wall. The most attractive part of the whole complex is the graceful Renaissance loggia at the west end of the garden, built by Giovanni Pieroni. It is decorated with stucco and frescos depicting the Trojan War by Baccio Bianco. Bianco also pandered to Wallenstein's taste for self-glorification by painting a fresco inside the palace showing his employer as the god Mars gliding above the clouds in his victory chariot.

The building's main façade occupies a whole side of Valdštejnské náměstí, but its serried banks of windows and three storeys are more formidable than pleasing. The overall design, which mingles late Renaissance and early baroque features, seldom achieves the harmonious elegance of later baroque architecture in Prague.

Niccolo Sebregondi laid out the gardens, at the far end of which is a large baroque riding school. The garden was ornamented with bronzes by the Duch artist Adrian de Vries and has a grotto with tufa stalactites at its north end (see page 128).

The palace belongs to the Ministry of Culture and contains the Jan Comenius Museum (see page 82).

Valdštejnské náměstí 4, Malá Strana. Trams 12, 22 and 18 and metro to Malostranská.

Patrician and Burgher Houses

In the Middle Ages house owners in Prague began the practice of identifying their properties by means of ornamental symbols, wall paintings or simply everyday objects attached to the façade. Trade premises would also be appropriately decorated (for example the violin-maker's house on Nerudova boasts a relief of three violins; and above the door of a former tavern on Husova two figures are shown carrying a vast bunch of grapes on a pole). In the 18th century Maria Theresa introduced the so-called *Conscriptionsnummern* (conscription numbers) for houses, not only to rationalise the addresses of city dwellers, but also to facilitate the systematic checking of army conscription lists.

Despite the introduction of house numbers many picturesque signs still survive. Most date from baroque times, although some go back even earlier. The houses described below (five out of many such) have retained their decorative idiosyncrasies. Only those that are restaurants or partly shops may be visited inside.

ROTTŮV DŮM (Rott House)
The cellars of this house still have their Gothic vaulting, and are themselves the successors to the foundations of a Romanesque dwelling on this site. In the 15th century a printshop was situated here which produced Prague's first printed bible in 1488.

The present neo-Renaissance aspect of the building is the result of alterations made in 1890, when it was owned by an iron merchant named Rott (his name is blazoned across the façade). The entire front wall was painted with decorative foliage and vignettes by Mikoláš Aleš. Notable are the charming emblematic figures for agriculture and the crafts whose iron tools were Rott's stock in trade.
Malé náměstí 3, Staré Město. Trams 17 and 18 and metro to Staroměstská.

Many of Prague's houses go back to baroque times or even further

The front of the Rottův Dům is decorated with murals and cartoons by Mikoláš Aleš

U DVOU ZLATÝCH MEDVĚDŮ
('At the Two Golden Bears')
The Renaissance portal of this house has been preserved, but otherwise its aspect is neo-classical, following rebuilding in 1800. Above the lintel two stone bears (originally gilded) can be seen. Animals (apparently often chosen at random) were very popular motifs for house signs.
Kožná ulice 1, Staré Město. Metro to Můstek.

U SAMUELA ('At Samuels')
This originally Gothic, later 'baroquised', house is situated in the pedestrian zone just below Václavské náměstí (Wenceslas Square). It takes its name from the relief on the corner of the building that shows the biblical King David as a child being anointed as future leader of his people by the prophet Samuel.
Na můstku 4, Staré Město. Metro to Můstek.

U TŘÍ PŠTROSŮ
('At the Three Ostriches')
The sign for this house was an advertisement: the 17th-century owner traded in ostrich feathers! Baroque alterations to the building were made in 1657, when the gables were added. An Armenian named Deodatus Damajan may have founded Prague's first coffee-house here in the 18th century, and long before that there was a restaurant on the ground floor (as there still is). In 1976 a luxury hotel was added to the facilities.
Dražického náměstí 12. Malá Strana. Trams 12 and 22 to Malostranské náměstí.

U ZLATÉHO JELENA
('At the Golden Stag')
The 'golden stag' here is a spectacular sculptural group above the portals, the work of Ferdinand Brokoff. It shows St Hubert kneeling before an amiable-looking deer: according to the legend St Hubert went hunting on Good Friday and encountered a stag with a golden crucifix in its antlers. This was taken as a warning that he should repent of his sacrilegious disregard of a holy day. The baroque house was built by Kilian Ignác Dientzenhofer in 1726.
Tomášská ulice 4, Malá Strana. Trams 12 and 22 to Malostranské náměstí; metro to Malostranská.

BAROQUE PRAGUE

In the 17th and 18th centuries the face of Prague changed, reflecting the triumph of the Habsburg dynasty and of Catholicism in Bohemia. The baroque style was the official stamp placed on the city by the victors in the struggle for the nation's body and soul.

Baroque architecture still dominates the historic areas of Prague, especially Malá Strana and Hradčanské náměstí. Here the nobility built their fabulous palaces, competing with each other and the Emperor in displays of wealth and splendour. Baroque forms were imported from Italy and at first chiefly Italian architects were employed.

The Catholic nobility's power and privilege was underlined by the allegorical sculptures of mythological heroes that ornamented the façades and gardens of their palaces. Atlas figures support their splendid entrances and representations of Hercules triumphant against his foes embellish many a stairway or avenue.

Similarly the magnificent baroque churches are symbols of the Catholic church's triumphant Counter-Reformation that followed the defeat of the Protestant cause at the Battle of the White Mountain in 1620.

Prague produced some great Baroque masters, although they were nearly all of foreign origin (František Kaňka is one notable exception). Christoph Dientzenhofer and his son Kilián Ignác were Germans, Jean-Baptiste Mathey came from France and Giovanni Santini-Aichel from Italy. The prolific Dientzenhofers built the loveliest of Prague's baroque churches – including the two St Nicholases on Malostranské náměstí and

Baroque doorways: a statement of power

Prague's houses and churches are rich in baroque decoration

Staroměstské náměstí – as well as many secular buildings. Catholic orders (especially the Jesuits) poured money into religious architecture and dozens of Gothic churches up and down the land were 'baroquised'.

Cupolas, towers, terraced gardens, pathos-ridden gesticulating statues – all these conjure a vision of a city once suffused with religious fervour, but simultaneously the playground of overmighty princes.

NÁMĚSTÍ JANA PALACHA (Jan Palach Square)

The square commemorates the student who burned himself to death on Wenceslas Square following the Soviet invasion of 1968. Palach was a 21-year-old student at the Faculty of Philosophy, which runs along one side of the square. A bust of him has been erected on the corner of the faculty building.

Trams 17 and 18 or metro to Staroměstská.

PETŘÍN (Petřín Hill)

Petřín Hill, an eastern outrider of the White Mountain, was made into a public park in the 19th century. In 1901 it was linked to the Kinsky Gardens by means of a gap in the Hunger Wall (see opposite).

According to the 11th-century chronicler, Cosmas of Prague, it derives its name from the Latin *petrus* ('stone') – the city did once get most of its building material from here. Later it was covered in vineyards and is today a wooded park with winding paths, much favoured by the Praguers in spring when its orchards are in blossom.

Lanová Dráha (Funicular Railway)

The funicular carries you up to the sights of Petřín. On the way it stops at the Nebozízek café and restaurant, which takes its name from the vineyard originally here; its terrace offers good views of the city.

Trams 6, 9, 12 and 22 to Újezd.

Petřínská Rozhledna (Petřín Tower)

At the top of the hill is a scaled down copy of the Eiffel Tower in Paris, made for the Prague Industrial Exhibition of 1891. There are 226 steps to the viewing platform and no lift, but the view is worth it – they say you can see the Alps on exceptionally clear days!

Open: May to September, 9am–11pm. Admission charge.

Jan Palach's sacrifice has not been forgotten

Kostel Sv Vavřince na Petříne (The Church of St Lawrence on Petřín)
The Germans call Petřín 'Laurenziberg' after the dedicatee of this church whose Romanesque forebear is first mentioned in records of 1135. Kilián Ignác Dientzenhofer and Ignác Palliardi rebuilt it in baroque style between 1735 and 1770. St Adalbert, the martyred 10th-century Bishop of Prague, is commemorated by a statue, an altar painting and a ceiling fresco depicting him founding the church.

Near by are stations of the cross culminating in a Calvary Chapel with 19th-century sgraffiti by Mikuláš Aleš. The church may be difficult of access during renovation.

Bludiště (The 'Mirror Maze')
An odd-looking pavilion near St Lawrence contains a labyrinth of mirrors. At the end of this is a diorama of the Prague students resisting the onslaught of the Swedes in the invasion of 1648.
Open: April to October, 9am–6pm. Admission charge.

Hvězdárna Hlavního Města (The People's Observatory)
The Astronomical Institute of the Czech Academy of Sciences is based here, but amateurs are allowed to look through the telescopes. The most modern is a 40cm Zeiss, but the older instruments, affectionately known as 'The King' and 'The Comet Finder', are still in use.
The Observatory's opening times are incredibly complicated. It is best to ring 2451 0709 if you want to arrange a visit.

Hladová Zed (Hunger Wall)
Charles IV had this great fortification built between 1360 and 1362, supposedly as a job creation scheme for the starving unemployed of the city (hence the name). The wall runs down the hill from the border of the gardens of Strahovský klášter (Strahov Monastery) in the northwest to that of the Kinsky Garden in the southeast.

The Calvary Chapel nestles among the wooded slopes of Petřín Hill

Access to Petřin Hill: by funicular from the station above Újezd (trams 12 and 22); from the gardens of Strahov Monastery (tram 22 to Památník písemnictví); from Vlašská ulice (trams 12 and 22 to Malostranské náměstí).

Obecní Dům

(The Community House)

By the last decade of the 19th century the increasing self-confidence of the Czech nation found expression in architecture. In Bohemia 'Community' or 'National' houses sprang up, combining representational, recreational and social facilities.

Naturally Prague had the biggest and best Community house; its full name – *Representační dům hlavního města Praha* (The Central Representation House of the City of Prague) – indicates its aspirations. It was to be the focus of the Czech capital, celebrating the Czech people. It was here that MPs issued the Epiphany Declaration of January 1918 demanding the setting up of an independent Czechoslovak state.

The site chosen for this display of civic and national pride was very appropriate; it was to be built where the court of the Jagiello king had stood, next to the 'Powder Tower' (see page 107). When the tower was built in the 15th century, the municipality had financed that too, and the burghers had inscribed a declaration on it stressing that it was erected 'to the honour ... of the citizens of the town'; their rulers were expected to get the message. The tower's main architect, Matěj Rejsek, has been remembered with a statue on the nearest corner of the Community House. A competition for the design was won by Antonín Balšánek and Oswald Polívka. Their project was realised in art nouveau style (see pages 28–9) between 1905 and 1911. It was the most ambitious art nouveau building in the whole country.

The exterior

The façade (currently in the process of much needed restoration) is a glorious clutter of glass ornament, wrought-iron railings and theatrical statuary. It is topped by a glazed dome, beneath which a huge arched gable frames Karel Spillar's symbolic mosaic 'Homage to Prague'. This is flanked by pathos-ridden allegorical sculpture representing 'The Humiliation' and 'The Rebirth of the Czech Nation'.

The interior

All the leading artists of the Czech Secession had a hand in the decoration of the marvellous interior. Immediately to your right is a vast restaurant with lavish décor: ornamental stucco and

Art nouveau in the ascendant

The wall paintings in the Smetana Hall, Obecní Dům, symbolise the dramatic arts

huge gilt chandeliers set off the acres of mural with themes such as 'Hop Growing', 'Viticulture' or 'Prague Welcomes Its Visitors'. Across the vestibule to your left is the equally ornate café, dominated at one end by a fountain with a nymph sculpted from Carrara marble by Josef Pekárek.

In the basement are a beer cellar and wine bar with murals by Mikoláš Aleš and ceramics by Jakob Obrovský and others.

First floor

The focal point is the **Smetanova síň** (Smetana Hall) where the composer's symphonic poem *Má Vlast* (My Country) is performed at the beginning of the Prague Spring Festival (see page 155). The sculptural groups flanking the stage represent Dvořák's *Bohemian Dances* and scenes from Smetana's opera *Vyšehrad*.

The most important of the other rooms is the **Sál Primátorský**, the circular mayoral hall decorated with paintings by Alfons Mucha, an artist who designed a pavilion for the Paris World Exhibition of 1900. Also impressive is the **Rieger sál** with Max Švabinský's large painted panels entitled *Prague Spring*, which feature portraits of leading Czech cultural figures.

Náměstí Republiky 5. Metro to Náměstí Republiky.

The restaurants are open 11am to 11pm, the café from 7am to 11pm. The wine and beer cellar may sometimes be reserved for travel groups. Inquire about tours of the upstairs rooms (which are not normally open) at the ticket office in the lobby or at PIS information bureaux.

Dientzenhofer built the splendid baroque Portheimka for his family

PALÁC JIŘÍHO Z PODĚBRAD (Podiebrad Palace)

Few remnants of Romanesque dwellings remain extant in Europe, but Prague has more than its fair share. The most striking is the Podiebrad Palace, the basement of which retains its original Romanesque cross-vaulting and fireplaces (see page 119). What now appears to be the cellar was once the ground floor of the house, before street levels were raised during flood protection works in the second half of the 13th century.

The palace was built at the end of the 12th century or the beginning of the 13th for the Lords of Kunštát and Podiebrad. In 1406, according to the city records, the owner was Lord Boczko of Kunštát, the uncle of George of Podiebrad (see box opposite), who inherited the building and lived here between 1453 and 1458.

In 1970 the restored palace was opened to the public. Inside it contains a small display of ceramics and one concerning the life of George of Podiebrad.

Řetězová ulice 3. Open: May to October, Tuesday to Sunday 10am–noon, 1–6pm. Admission charge. Metro to Můstek.

LETOHRÁDEK PORTHEIMKA (Portheimka Summer Palace)

Notwithstanding its shamefully dilapidated state, the lightness and grace of this diminutive baroque *Lustschloss* shines through. The architect Kilián Ignác Dientzenhofer built it for his own family in 1729. It derives its present name from the industrialist who bought it in the 19th century. An oval saloon with frescoed ceiling looks out on what was once a pleasant garden.

Matoušova ulice 9, Smíchov. Metro to Anděl.

PRAŠNÁ BRÁNA (Powder Tower)
King Vladislav Jagiello laid the foundation stone for this huge gate-tower in 1475. It replaced an earlier gateway at the point where the trade from the east entered the city. The new fortification was modelled on the Staré Město tower at the end of the Charles Bridge (see pages 68–9). For a while the king had his residence next door to it, but later the hostility of the Hussite burghers compelled him to retreat to the Hradčany.

In the 18th century the tower was used as a powder magazine and was badly damaged when Frederick the Great attacked the city in 1757. Josef Mocker conscientiously 're-Gothicised' it in 1875, when the statues were placed on the façade.

The tower can be climbed at weekends by those prepared to negotiate the 186 steps for a good view of Staré Město.
Náměstí Republiky. Open: May to September, Saturday, Sunday and Wednesday 10am–6pm; April and October, 10am–5pm. Admission charge. Metro to Náměstí Republiky.

JIŘÍ Z PODĚBRAD (George of Podiebrad – King of Bohemia 1457–71)
After the religious reformer Jan Hus was burned for heresy (see page 37) his followers in Bohemia achieved considerable military success against the Catholic powers. However, they were themselves divided between radical 'Táborites' (Tábor was the biblical name they gave to the place where they formed their movement) and moderate utraquists (so-called because they believed Communion should be administered by the laity in both bread and wine).

The greatest leader of the Utraquist faction was George Podiebrad, who became regent during the minority of King Ladislav. When the latter died of the plague in 1457, Podiebrad was elected king by the Bohemian Diet. To mark this event a chalice (the Utraquist symbol) was placed on the façade of the Týn Church (see page 110).

The reign of George of Podiebrad marked the last, albeit glorious, phase of Hussite independence: the king withstood the machinations of the papacy and tried to unite the princes of Europe against the ever-increasing Turkish threat. He was the most able leader of his day and is justly regarded as one of the greatest heroes of Bohemian history.

Ornamented window, Portheimka

Staroměstské Náměstí

(Old Town Square)

If Hradčany was the centre of royal authority in Prague, Staroměstské náměstí was the focus of people power. John of Luxembourg first gave permission for a Town Hall to be built here in 1338, but an independent-minded municipality had existed long before that.

The square had been the focus of trade and exchange in the 12th and 13th centuries. Goods came in from the east through the customs house in Týn Court (see page 113) and out on the ancient trade route that crossed the Vltava via the old Judith Bridge (see page 68). Romanesque remains show there was a thriving community then. Most of these buildings were rebuilt in the Gothic style, later receiving a Renaissance or baroque cladding).

The astronomical clock in Old Town Square

Staroměstské náměstí has always been at the heart of Czech identity. After the first defenestration of Prague in 1419 (see page 123), the ringleader, Jan Želivský, was executed on the square. The rebellious Protestant nobles met the same fate after the Battle of the White Mountain (1620 – they are recalled by 27 white crosses set in the paving in front of the Town Hall.) The Hussites (see page 37) had their hour of glory when their candidate, George of Podiebrad, was elected king in the Town Hall in 1458. Jan Hus himself is honoured with a massive symbolic statue.

In modern times Staroměstské náměstí has again been the setting where history was made. It saw the jubilation that marked the beginning of Communist rule in 1948, the furious protest at its continuing usurpation of power in 1968, and joy at its demise in 1989.

STAROMĚSTSKÁ RADNICE (Old Town Hall)

At the southwest corner is the Old Town Hall, which has reached its present proportions by steadily swallowing up neighbouring buildings between the 14th and 19th centuries. In the Great Hall are Václav Brožík's heroic representations of the *Election of George of Podiebrad* and *Jan Hus before the Council of Constance*, complementing the patriotic mosaics in

The baroque Church of St Nicholas presides over the entire district

the entrance hall. On the southern façade is the famous **Orloj** (astronomical clock). On the hour every hour Christ and the 12 apostles emerge and the skeleton of Death tolls a bell with one hand while holding a sand-glass in the other. You can also see a turbaned Turk, a miser and a vain man admiring himself in a mirror. The complicated clockface shows everything from hours and days to equinoxes and phases of the moon.

According to legend, after the astronomer Master Hanuš put the finishing touches to the clock in 1490 the municipality had him blinded so that he should not repeat his achievement elsewhere. Hanuš groped his way up the clock tower and ruined the mechanism, which refused to function for 80 years.

Open: March to October, Thursday to Sunday 9am–6pm; November to February, 9am–5pm. Guided tours of the interior every hour. Admission charge. Restoration may limit access.

Staroměstské náměstí is in a pedestrian zone. It is approached by trams 17 and 18 and the metro to Staromestská, followed by a short walk along Kaprová.

KOSTEL SVATÉHO MIKULÁŠE
(Church of St Nicholas)

At the northwest corner of the square stands Kilián Ignác Dientzenhofer's marvellously proportioned baroque church, with its cool white façade, built for the Benedictines between 1732 and 1735. The interior is less sumptuous than one might expect from the exterior – perhaps because much was removed when Joseph II turned it into a warehouse in the late 18th century. Since 1920 it has belonged to the refounded Czech Hussite Church.

Open: 10am–noon, 2–5pm, and evenings during concerts. Closed: Monday and Saturday.

The distinguished Kinský Palace, designed by Kilián Ignác Dientzenhofer

PALÁC GOLTZ KINSKÝCH (Kinský or Golz-Kinský Palace)

On the east side of the square is the Kinský Palace of 1765. Franz Kafka attended a school in the palace and his father kept a haberdashery shop on the ground floor. At present an exhibition hall for the graphics collection of the National Gallery, it has been acquired by the Kafka Society who plan to turn it into a library and museum.

Staroměstské náměstí 12. Tel: 2481 0758. Open: Tuesday to Sunday 10am–6pm.

KOSTEL PANNY MARIE PŘED TÝNEM (The Church of Our Lady before Týn)

South of the Kinský Palace is the Venetian style old parish school (Týnská škola). Through the third arch from the left access is gained to the Týn Church, built by Petr Parléř's workshop in the 14th century. It was once the stronghold of the moderate Hussites known as Utraquists (see page 107). Their symbol, a huge gilded chalice, hung on the façade until replaced by a Counter-Reformation image of the Virgin Mary (embellished with gold from the chalice).

The bleak interior contains as highlights baroque paintings over the high altar by Karel Škréta, a medieval *pietà* in the side-chapel at the east end, and the tomb of Tycho Brahe, on the fourth pier to the right.

Access is limited, but usually possible between 4pm and 5pm daily, or for mass at 6pm.

U KAMENNÉHO ZVONU (The House of the Stone Bell)
The existence of this remarkable Gothic house was unknown until the 1960s, when a survey revealed the medieval treasure encased in a baroque shell. It is thought to have been a palace for Queen Elizabeth, wife of John of Luxembourg.
Staroměstské náměstí 16. Tel: 2481 0036. Open: 10am–6pm and sometimes for evening concerts on the upper floor.

DŮM U MINUTY (House of the Minute)
The most striking aspect of this house, rebuilt in the 17th century in the style of the Lombardy Renaissance, is that its walls are completely covered with sgraffiti of mythological and biblical scenes.
Staroměstské náměstí 2. Not open to the public.

U ZLATÉHO JEDNOROŽCE (The House at the Golden Unicorn)
This building reveals several layers of architecture – Romanesque, Gothic, Renaissance, and finally a baroque façade. Bedřich Smetana founded a music school here in 1848.
Staroměstské náměstí 20. Not open to the public.

U MODRÉ HVĚZDY (House at the Blue Star)
This historic inn has a Romanesque hall below ground, while the brick arcades are Gothic. The *vinárna* (wine bar) here, known as U Bindrů, has been in continuous operation since the 16th century. Local characters live on in the names of its main dishes which bear such names as 'Hanuš' (of astronomical clock fame) and 'Mydlář' (the executioner of the Protestant nobles in 1621, much admired for his cool way with an axe).
Staroměştské náměstí 25. Tel. 2422 7541. Open: daily 11am–midnight.

POMNÍK JANA HUSA (The Jan Hus Monument)
On 6 July, 1915, the 500th anniversary of the judicial murder of Jan Hus, this vast art nouveau monument was unveiled. Ladislav Šaloun's sculpture exudes pathos and patriotism. Hus stands grimly, flanked by 'the defeated' and 'the defiant'. The inscription runs: 'Truth will prevail' – a quotation from Hus's preaching. The monument became a rallying point and emblem of Czech patriotic feeling. For that reason the Nazis covered it with swastikas in 1939; in 1968, on the other hand, it was draped in black cloth, a sign of mourning for Czech independence crushed by the Russians.

House at the Golden Unicorn

Theological Hall in the Strahov Monastery, with its magnificent ceiling frescos

STRAHOVSKÝ KLÁŠTER (Strahov Monastery)

Strahov owes its name to its commanding position on Charles IV's city fortification (*strahovní* means 'to watch over'). The monastery (see page 133) is approached either from Pohorelec no 8, up some narrow steps, or through a baroque archway further east. The arch is crowned with J A Quittainer's statue of St Norbert, founder of the Premonstratensian Order that occupied the monastery after its foundation by Prince Vladislav II in 1140. In 1627 the monks acquired the remains of the saint.

In the courtyard you will pass the deconsecrated **Kostel svatého Rocha** (Church of St Roch) erected by Rudolf II in gratitude to the saint for his efforts in ensuring that Prague was spared the 1599 plague epidemic. Beyond is the **Kostel Nanebevzetí Panny Marie** (Church of the Assumption). Quittainer's impressive *Immaculata* over the portal provides a foretaste of the rich baroque interior. Mozart twice played on the church's mighty organ which has 4,000 pipes and 63 stops.

Close by is a ticket office for visits to Strahov's most spectacular sights, the **Teologický sál** (Theological Hall) and the **Filosofický sál** (Philosophical Hall), both libraries. The Theological Hall was built in 1671 with a pleasing barrelled vault by Giovanni Orsi. One of the monks, Siard Nosecký, painted frescos illustrating man's struggle to acquire wisdom. A similar theme forms the leitmotif of the neighbouring

Philosophical Hall, where the ceiling was painted by the great Viennese-trained master, Franz Anton Maulpertsch in 1784. His *Struggle of Mankind to Know Real Wisdom* boldly includes pre-Christian figures, such as Alexander the Great, Aristotle, Plato, Socrates and Diogenes the Cynic (sitting in his barrel)
Pohořelec 8, Hradčany. Tel: 2451 0355. Open: Tuesday to Sunday 9am–noon, 1–5pm. Tram 22 to Památník písemnictví. Strahov also houses the Památník Národního Písemnictví (Museum of Czech Literature – see page 81).

TROJSKÝ ZÁMEK (Troja Palace)
The red painted Troja Palace on the outskirts of Prague is a splendid piece of triumphalist architecture commissioned by Habsburg loyalist Count Wenceslas Šternberk just after the spectacular defeat of the Turkish armies at the gates of Vienna in 1683. Jean-Baptiste Mathey designed the summer palace with a formal French garden.

Baroque ornamentation on the Troja Palace

The grand steps on the garden side are embellished with statues representing the battle between Gods and Titans. In the Great Hall, busts of Habsburg emperors line the wall in the Roman manner and a huge mural by Dutch artist Abraham Godin shows the triumphal procession of Leopold I after the Turks had been put to flight.

The palace is now administered by the Museum of the City of Prague and contains a good display of Czech 19th-century painting.
U trojského zámku 1, Troja. Tel: 6641 5144. Open: April to October, Tuesday to Sunday 10am–6 pm. Gardens open: Tuesday to Sunday 10am–5pm all year. Admission charge for palace. Metro to Nádraží Holešovické, then bus 112.

TYLOVO DIVADLO (Tyl Theatre)
This delightful neo-classical theatre is named after the Czech dramatist, Josef Týl, but is also known as the Stavovské divadlo (Estates Theatre), as it belonged to the Bohemian Diet in the early 19th century. Mozart's *Don Giovanni* and *La Clemenza di Tito* were premièred here in 1787 and 1791 respectively.
Ovocný trh 6. Tel: 2421 4339. Performances usually begin at 7 pm.

TÝNSKÝ DVŮR (Týn Court)
Týn Court was also hnown as *Ungelt* because money (*Geld*) was exchanged and customs dues paid here from medieval times up to 1773. The word *týn*, (an enclosed area or courtyard) gave the whole complex of buildings its name and was extended also to the neighbouring church and school (see page 110). Part of Týn Court is being converted into luxury hotel suites.
Immediately northeast of Staroměstské náměstí. Metro to Můstek.

VÁCLAVSKÉ NÁMĚSTÍ (Wenceslas Square)

More of a boulevard than a square, this is the best known part of Prague, although not the most attractive. The area was once a horse market but has long been the focus for political demonstration. The most dramatic events of the Velvet Revolution took place here, including the historic moment when Václav Havel and Alexander Dubček addressed the crowd from the balcony of the offices of the Socialist Party newspaper (no 36).

The southeast end of the square is dominated by the **National Museum** (see page 84); below that is Josef Myslbek's monumental equestrian statue of St Wenceslas (1912), a constant rallying point for protest or jubilation. It was here that the student Jan Palach set fire to himself on 16 January 1969 in protest at the Russian invasion. In front of the monument is a small shrine to the victims of Communist repression.

The buildings

The square is a showcase for 19th- and 20th-century public buildings. There are fine art nouveau edifices (the Peterkův dům and the Evropa Hotel – see pages 28–9), imposing neo-Renaissance blocks and some of Prague's better modern architecture, such as Bata shoe store (Dům obuvy, no 6), now restored to family ownership and newly modernised. The Hotel Juliš at no 22 was designed by Pavel Janák and (like the shoe store) is a so-called 'Constructivist' work from the 1920s. Worth a glance also is the leisure complex of the Lucerna Palace (no 61) built by Václav Havel's grandfather. The Wiehl House at no 34 is a colourful example of Bohemian neo-Renaissance.

The area encompassing Václavské náměsti, Národní and Na příkopě is the commercial centre of Prague and known as the Golden Cross. Approach is by metro to Muzeum or Můstek.

Nightlife in Wenceslas Square

Vltava Bridges

The Prague section of the Vltava is spanned by 15 bridges (*mosty*). The oldest, Karlův most (Charles Bridge – see pages 68–9) was also the only one up until 1836.

The great era for bridge-building was the late 19th and early 20th century, when the Palacký, Mánes, Čech and Hlávka bridges were constructed. The **Hlávkův most** was named after a rich contractor who financed its construction. Notable are its two allegorical sculptures of *Humanity* and *Work* by Jan Štursa. The **Čechův most**, named after the romantic poet Svatopluk Čech, has attractive turn-of-the-century decoration and lamp-posts. The historian František Palacký, the father of the Czech national revival, is honoured in the name of the **Palackého most**, a bridge crossed by Albert Einstein twice daily as he walked to and from the university.

The **Most Legií** was inaugurated by Emperor Franz Joseph in 1901, but its name commemorates the Czech legions who fought against his crumbling empire in the World War I. The **Mánesův most** takes its name from the patriotic painter Josef Mánes (see page 90) and a monument to him stands at the Josefov end.

Hlávkův most – trams 3 and 8; metro to Vltavská.
Čechův most – tram 17.
Palackého most – metro to Karlovo náměstí.
Most Legií – trams, 6, 9, 12, 18 and 22.
Mánesův most – trams 12, 17, 18 and 22.
Bránický most – trams 3, 17 and 21.

An elaborate lampstand base on the Legií (Legion) Bridge

Bridges seem to be regarded as political barometers and to change their names accordingly. The Mánes Bridge was originally dedicated to the Austrian heir to the throne, Archduke Franz Ferdinand and the Bridge of the Legions became the Bridge of the First of May under the Communists. Poignantly, the unofficial name for the **Bránický most** is the 'Bridge of the Intelligentsia', because intellectuals were used as forced labour to build it for the Stalinist regime of the 1950s.

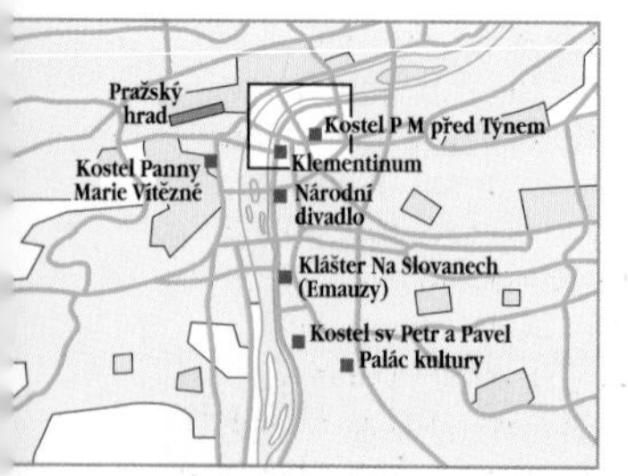

The Heart of the Old Town

This walk through Prague's historic Staré Město (Old Town) offers a kaleidoscopic tour of the Czech core of the city. ***Allow about 1½ hours.***

The point of departure is Staroměstské náměstí (Old Town Square), which is reached on foot from the Staroměstská metro station (line A) or tram stop (nos 17 and 18).

1 OLD TOWN SQUARE

In the summer months Old Town Square (see pages 108–11) is an ongoing spectacle where you can encounter anything from living sculpture to Peruvian Indians playing reed flutes. There are reminders of a Hussite past in the great Kostel Panny Marie před Týnem (Týn Church) at the east end and the

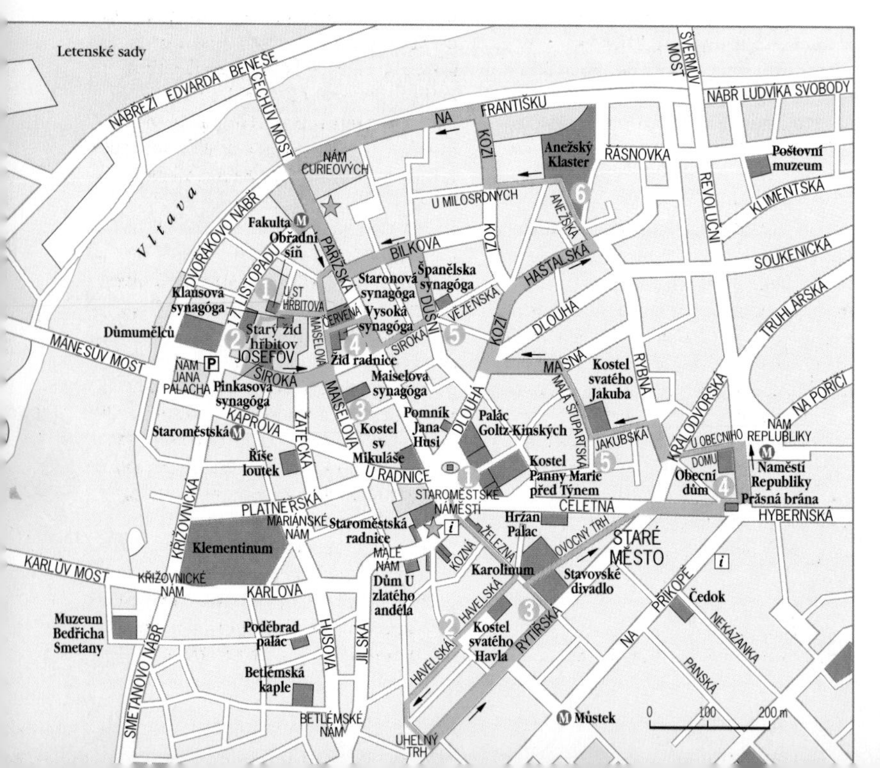

modern monument to Jan Hus dominating the northern half of the square. Near by, Protestant nobles were barbarously executed after rebelling against the Habsburgs. In the more recent past, the 1948 Communist takeover was announced by Klement Gottwald to cheering crowds from a balcony of the Palác Goltz-Kinských (Kinsky Palace) on the east side. A later generation threw Molotov cocktails as Soviet tanks advanced into the square in 1968.

Leave the square by Železná street and turn right into Havelská.

2 HAVELSKÁ

On your left is Kostel svatého Havla (St Gall's Church), where early campaigners preached against the abuses of the Catholic hierarchy. Outside, a lively market sells everything from fruit and toilet paper to figurines of the good soldier Švejk (see page 7).

Turn left out of Havelská across Uhelný trh, then left into Rytířská.

3 RYTÍŘSKÁ

At the end of the street is the graceful Stavovské divadlo (Estates Theatre), which Miloš Forman chose for scenes from the film *Amadeus*, since it remains exactly as it was on *Don Giovanni*'s first night. As you walk down Ovocný trh beside the theatre, you will see an ornate Gothic window protruding from the wall of the Karolinum (Charles University, see page 66). The street ends in Celetná; on the Cubist corner building (no 34) note the petite Black Madonna, a survivor from an earlier baroque building here.

4 OBECNÍ DŮM

A right turn into Celetná brings you under the medieval Prašná brána (Powder Tower) and then left, on into náměstí Republiky, which is dominated by the art nouveau Obecní dům (Community House, see pages 104–5). Its florid statuary, obsessional geometrical motifs and general air of faded glory can be savoured in the ground-floor café. The independence of 'Czechoslovakia' was proclaimed here in 1918.

Turn left up U Obecního domu, right up Rybná and left up Jakubská, passing along the side of Kostel svatého Jakuba (St James's Church), whose entrance is in Malá Štupartská.

5 KOSTEL SVATÉHO JAKUBA

One of the loveliest of Prague's baroque churches (see page 37), St James is beloved by the locals for its sung masses on Sundays, not least because of its superb acoustics. It has stunning frescos and an exuberant stucco façade.

At the northern end of Malá Štupartská turn left into Masná, cross Dlouhá and enter Kozí, bearing right into Haštalská and then left into Anežská.

6 ANEŽSKÁ

At no 3 is the Vinotheka Anežská, where you can stop off to enjoy the excellent wines from the vineyards of Velké Žernosky. The Müller Thurgau (as the proprietress will assure you) was chosen by President Havel for state banquets. At the end of the street the Anežsky klášter (St Agnes Convent – see pages 26–7) is entered through a gate in the wall. Its fine Gothic architecture and picture gallery are worth a leisurely hour.

From St Agnes head for Na Františku, via U Milosronych and Kozí. It is then a short walk westwards along Na Františku to Pářižská and the trams at Pravnicka Fakulta.

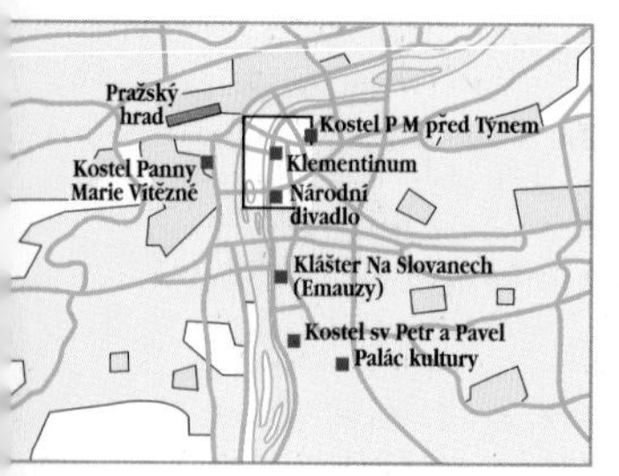

Národní Třída to Křížovnické Náměstí

This route encompasses places associated with a vanishing Habsburg, an aspiring playwright, aggressive agitators and astute propagandists. *Allow 1½ hours.*

Begin at Národní divadlo (National Theatre) reached by trams 6, 9, 18 and 22.

1 NÁRODNÍ DIVADLO

The National Theatre (see page 85) opened in 1881 with a performance of Smetana's patriotic opera *Libuše*. Opposite it, on the corner of Národní and the embankment, is the Café Slavia. In the inter-war years this was the haunt of literati including Nobel Laureate Jaroslav Seifert; under the Husák regime opposition writers met here to pass around their *samizdat* manuscripts.

A few steps to the north on the embankment brings you to the monument to the Habsburg Emperor Franz I. Its equestrian statue is missing: national feeling decreed that it should be banished to the National Museum. *Cross the square into Karolíny Světlé.*

2 KAPLE SVATÉHO KŘÍŽE

To your right is the Romanesque Kaple svatého Kříže – the Rotunda of the Holy Cross.

A detour north brings you towards Anenské náměstí and the Divadlo Na zábradlí (theatre) where Václav Havel began his underground career in the 1960s – as a stagehand. Turn right up Náprstkova, which brings you to Betlémské náměstí.

3 BETLÉMSKÁ KAPLE

The square takes its name from the

gabled Bethlehem Chapel (see page 36), built by pre-Hussite reformists who had been denied the right to build a church – which is why a 'chapel' could accommodate 3,000! The Communists encouraged the building of the present replica – apparently Hussites were regarded as ideologically sound.
Turn left into Husova.

4 HUSOVA

Shortly on the right is Kostel svatého Jiljí (St Giles's Church). A touching fresco inside shows the hermit stricken by an arrow fired by an archer of the Visigoth king. The arrow had been destined for St Giles's pet deer, which looks surprised but unharmed.

A detour to your left along Řetězová takes you to Poděbrad palác (House of the Lords of Kunštát – no 3) with a perfectly preserved Romanesque cellar (open: May to October, Tuesday to Sunday, 10am–noon, 1–6pm; admission charge – see page 104).

Retrace your steps. Continue along Husova to the junction with Karlova. A detour right can be made here to the picturesque Malé náměstí; otherwise turn left into Karlova. Note the imposing Clam-Gallas Palace (see page 94) just beyond the junction, with its portals supported by two Atlas figures.

5 KARLOVA

Karlova is studded with up-market antique shops. Shortly on the right is the Klementinum (Clementinum), a huge Jesuit seminary set up at the beginning of the Counter Reformation (see page 70). It was a centre of learning, as well as propaganda, and Johannes Kepler once scanned the night skies from its observatory.
Karlova opens into Křížovnické náměstí.

6 KŘÍŽOVNICKÉ NÁMĚSTÍ

Knights of the Cross Square (see page 71), has two churches – another Jesuit one and the Franciscan Church (see page 37) that belonged to the Knights of St John, guardians of the Judith Bridge (Charles Bridge's predecessor). Next to the bridge is a cast-iron statue of Charles IV.
From Křížovnické náměstí you can either stroll straight on to Malá Strana across the Karlův most (Charles Bridge) or turn right and right again to follow Platneřská to Mariánské náměstí and eventually Staroměstské náměstí. Alternatively, pick up tram 17 or 18, both of which run from the embankment near by.

Statue of Charles IV overlooking the bridge that bears his name

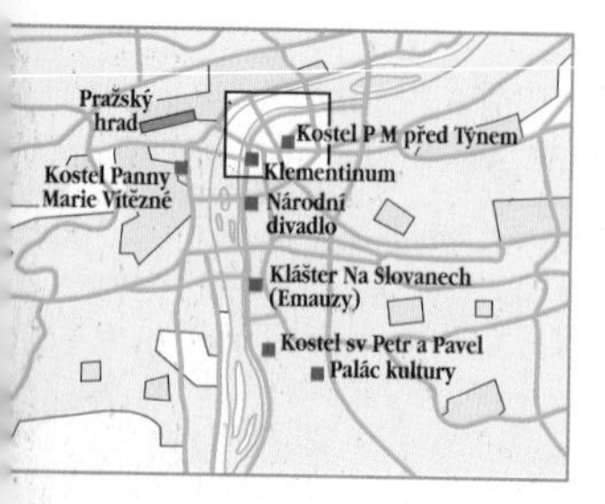

Josefov

This walk covers the former Jewish ghetto. In 1850 it acquired the name Josefov (Joseph's Town, see pages 60–3) in honour of the Emperor Joseph II whose Edict of Toleration in 1781 lifted many restrictions on Jews. See map on page 116 for route. *Allow about 1 hour.*

Take tram 17 to the Právnická Fakulta stop near Čechův most. Walk down Pařížská and bear right into Maiselova, then right again into U starého hřbitova.

1 U STARÉHO HŘBITOVA

Tickets can be purchased for all the sights of Josefov in the Klausová Synagóga (Klausen Synagogue) to your left at the end of the street. On the right is the Obřadní síň (Ceremonial Hall) which features pictures by children in the Terezín concentration camp. The Klausen Synagogue has a print display with items going back to 1512, the year when the first Jewish books to be printed anywhere were produced on this site. *Enter Starý židovský hřbitov (Old Jewish Cemetery) by the adjacent gate.*

2 STARÝ ŽIDOVSKÝ HŘBITOV

In Hebrew the name for this tranquil, yet haunting, place translates as 'The House of Life'; for over three centuries it was the only permitted burial place of the Prague Jews, so that some 12 layers of mortal remains had to be piled on top of one another (see page 63). The most visited tomb is that of the famous scholar Rabbi Löw. Leaving the cemetery on the south side you pass the Pinkasova synagóga (Pinkas Synagogue). Inside is a memorial to the Czech and Slovak victims of the holocaust, 77,297

Head coverings are offered at the Old-New Synagogue

names inscribed along a wall, all culled from the Nazis' pedanticaly precise transport files.

Turn left along Široká and second right back into Maiselova.

3 MAISELOVA SYNAGÓGA

On your left you soon come to the Maisel Synagogue, named after the ghetto's most famous mayor, who was also Rudolf II's minister of finance. Inside is a display of ritualistic objects, such as circumcision instruments and combs used in preparing the dead for burial.

Retrace your steps along Maiselova as far as the junction, on your right, with Červená.

4 ČERVENÁ

On your right is the diminutive rococo Židovská radnice (Jewish Town Hall). The cream and pink building's most notable feature is a Jewish clock on the front with a Hebrew dial, the hands of which move anti-clockwise (Hebrew script also goes from right to left). The Town Hall is the centre of the Jewish community and incorporates a kosher restaurant.

Directly opposite is the Staronová synagóga (Old-New Synagogue), one of the earliest in Europe, with stepped brick gables and a fine vaulted Gothic ceiling inside. A wrought-iron surround in the centre encloses the *bimah*, the lectern for readings from the Torah. Paper skullcaps must be purchased in the vestibule before entering the sacred area.

Adjoining the Town Hall is the Vysoká synagóga (High Synagogue), which is the main exhibition hall of the Jewish Museum. The abundance of Jewish mementoes and official buildings in Josefov is thanks to Hitler's decision to make this area into an official Nazi display, described as 'An Exotic Museum of an Extinct Race'.

The rococo elegance of the Jewish Town Hall

At the end of Červená turn right into Pařížská, and left into Široká, going as far as Vězeňská.

5 ŠPANĚLSKÁ SYNAGÓGA

The Spanish Synagogue on Vězeňská lives up to its name, with its Moorish appearance and Alhambra-like interior. The original synagogue here was founded by Sephardic Jews fleeing the inquisition.

Return to Pařížská and Čechův most via Dušní and Bílkova.

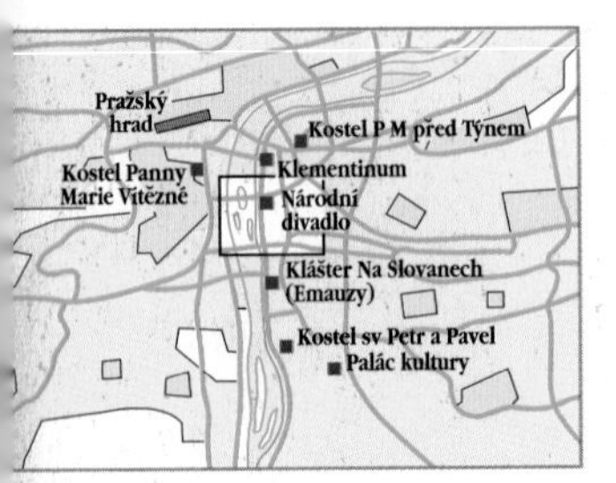

Nové Město

'New' is a decidedly relative term in Prague: the original Nové Město was founded by Charles IV as long ago as 1348. The king created this new quarter of the city in order to link Staré Město with the fortress of Vyšehrad. In the late 19th century the area was largely reconstituted, and a slum was transformed into streets and squares exuding middle-class respectability and self-confidence. *Allow 1 hour.*

Begin at Národní divadlo (trams 6, 9, 18 and 22) and walk east along Národní.

1 NÁRODNÍ DIVADLO

On your right is Kostel svaté Voršily (the Ursuline Church), part of a convent, which is usually open, so that one can drop in to admire the baroque frescos. On the outside is a striking statue of St John of Nepomuk; according to the legend, he was thrown into the Danube on 20 March 1393 on the orders of Wenceslas IV for refusing to reveal secrets of the Queen's confession.

Across the street are two exceptionally fine examples of art nouveau architecture, the Pojištovna Praha (Prague Savings

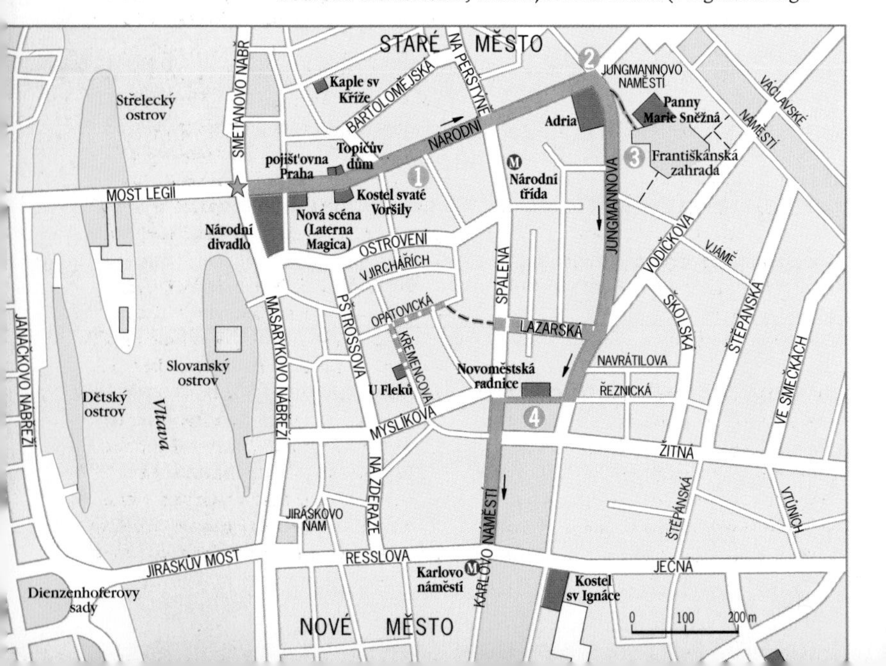

Bank) at no 7 and the Topičův dům (Topic publishing house) at no 9. A feature of the bank is the mosaic slogans above the windows advertising life insurance, loans, pensions and 'dowries'.

Further up the street on the right (no 16), a small shrine under an arcade commemorates the victims of police brutality on 17 November 1989, the so-called *masakr* (massacre) that triggered the 'Velvet Revolution'. Near by is the British Council building, recently refurbished but retaining a 'Constructivist' façade from the 1930s.

2 JUNGMANNOVO NÁMĚSTÍ

At the end of Národní is Jungmannovo náměstí, named after a leading figure of the 19th-century Czech literary revival, whose statue dominates the square. To the south is the Cubist Adria Palace, with a pleasant terrace café. The basement contains a theatre where Civic Forum used to meet after the 17 November 1989 'massacre'.

On the eastern edge of the square is the celebrated Cubist lamp-post, which stands close to the entrance to Chrám Panny Marie Snezná (Church of Our Lady of the Snows – see page 38). Charles IV planned this as a great coronation church, but the money ran out when only the chancel was complete. The soaring baroque high altar is impressive, but the whole has a rather gloomy atmosphere.

Bear left out of the church.

3 FRANTIŠKÁNSKÁ ZAHRADA

You soon come to the entrance to the Františkánská zahrada (Franciscan Gardens) to the southeast, part of the Franciscan monastery that ranges along their northwest wall. The gardens are a pleasant place to eat a fast food lunch

In Prague even the New Town Hall (scene of the first defenestration) is medieval

picked up from one of the purveyors on Václavské náměstí (Wenceslas Square) only a couple of minutes from here.

Returning to Jungmannovo náměstí, bear left down Jungmannova. Just beyond the junction with Vodičkova, Karlovo náměstí begins. On the way you can detour to U Fleků, where a famous stout is brewed on the premises (see page 170) Turn right into Lazarská, go through a passageway at Spálená 15 into Opatovická, then turn left for Křemencova 11. The beer hall is on your right. Retrace your steps to the end of Lazarská, then turn right into Vodičkova.

4 NOVOMESTSKÁ RADNICE

On your right is the ancient-looking Novoměstská radnice (New Town Hall); the tower, entrance hall and cellars are genuinely medieval. The first defenestration of Prague took place here in 1419 when Hussites threw Catholic councillors out of the windows for refusing to release reformist prisoners.

Trams back to the centre may be picked up on Ječná, which traverses Karlovo náměstí.

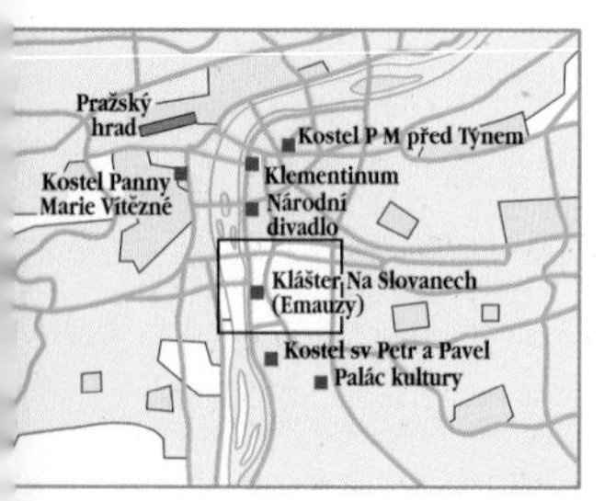

Churches of Nové Město

Classified as 'museums' under Communism, the many churches of Prague have begun to regain their pastoral functions and identity. This walk gives a taste of the rich heritage left by Prague's two great waves of Gothic and baroque church-building in the Nové Město (New Town). *Allow 2 hours.*

Begin at the metro station on Karlovo náměstí (line B) and walk west along Resslova.

1 KOSTEL SVATÉHO CYRILA A METODĚJE

On your right is the Church of St Cyril and St Methodius, the Czech Orthodox cathedral (see page 37). A plaque on the wall recalls that here the Free Czech paratroopers who had assassinated the Nazi governor of Bohemia, Richard Heydrich, made their last stand on 18 June 1942 (see page 142).

Across the street is Kostel svatého Václava (St Wenceslas Church). The Hussite congregation, which now owns St Wenceslas, is the modern successor to the warring Hussites

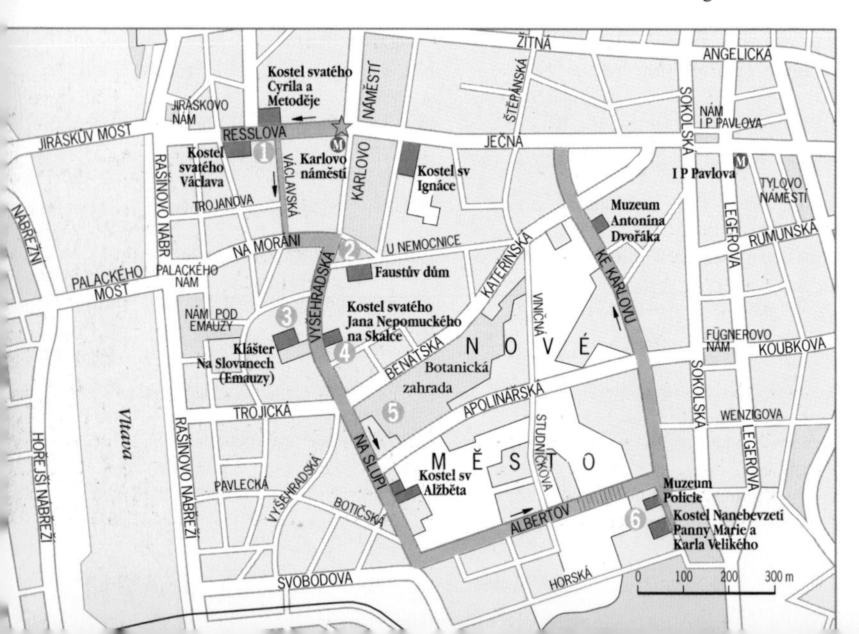

(see page 37) suppressed by the Counter-Reformation.
Walk back up Resslova, turn right into Václavská and go as far as Na Moráni. Turn left up to the junction with Vyšehradská, then right.

2 FAUSTŮV DŮM

On the corner with U nemocnice is the baroque Faustův Dům ('Faust House') where the opening scene of the Czech version of the Faust legend takes place. Rudolf II's English alchemist, Edward Kelley, is supposed to have lived in an earlier Renaissance house on this site.
Walk straight on.

3 KLÁSTER NA SLOVANECH (EMAUZY)

The Emmaus Monastery is one of the few buildings in Prague to suffer bomb damage in the war. The most dramatic aspect of its restoration are the two reinforced concrete sail-like spires (1968).
Continue further along on the left.

4 KOSTEL SVATÉHO JANA NEPOMUCKÉHO NA SKALCE

The Church of St John Nepomuk on the Rock (see page 38) is one of Kilián Dientzenhofer's loveliest baroque creations. It is difficult to access (try mass times) but you can still admire how the architect has cunningly turned to advantage the steep and narrow site on which his elegant little church perches.
As Vyšehradská merges into Na slupi you can turn off left into the Botanical Gardens.

5 BOTANICKÁ ZAHRADA

The glasshouse of exotic flora is worth the small entrance charge, and you can lose yourself happily for half an hour among the dense shrubbery and meandering paths (see page 42).
Continue along and turn left up Albertov. Climb the steps at the end on to Ke Karlovu.

6 KOSTEL NANEBEVZETÍ PANNY MARIE A KARLA VELIKÉHO

On your right is the 14th-century Church of the Assumption of Our Lady and Charlemagne, inspired by Charlemagne's burial chapel at Aachen. Because the stellar vaulting is built without central supports, the architect was accused of being in league with the devil – how else could he have pulled off such a feat? Remarkable features include 'holy steps' (which pilgrims had to ascend on their knees), theatrical baroque balcony scenes and a Bethlehem grotto in the crypt.

The Muzeum Policie (Police Museum) is adjacent to the church. You return along Ke Karlovu, passing the Muzeum Antonína Dvořáka (Antonín Dvořák Museum – Vila Amerika, see page 88) on the way.
Trams to the centre leave from Ječná at the far end of the street.

Display at the Muzeum Policie

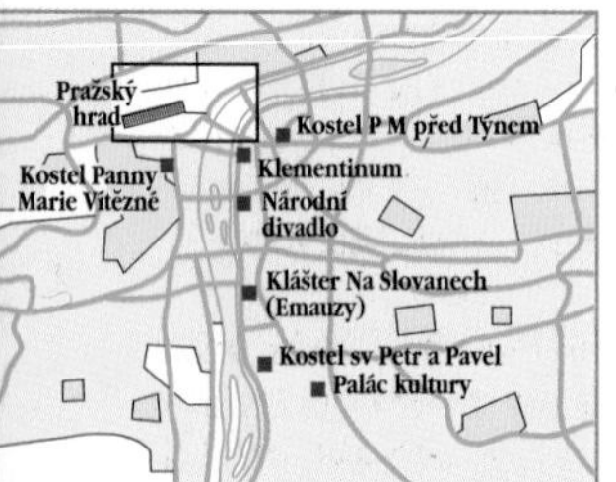

Letná Park and the Royal Gardens

This walk takes you above the left bank of the Vltava, away from the bustle and pollution of the city centre, allowing you to breathe the purer air of Letna and the Královska zahrada (Royal Gardens). *Allow 2 hours.*

Start from Čechův most (nearest tram stop: Právnická Fakulta, no 17).

1 ČECHŮV MOST

The bridge has an attractive turn-of-the-century design, with angels on columns at each end, sunburst-topped lampstands and art nouveau railings.

Cross the busy nábřeží Edvarda Beneše at the far side and climb the steep double flight of steps up to Letenské sady (Letná Park).

2 LETENSKÉ SADY

The vantage point on the top of the steps once boasted the largest Stalin monument in the Eastern Bloc (30m high). After Krushchev denounced Stalin, it was blown up. Where Stalin once stood there is now a gigantic 'metronome sculpture'

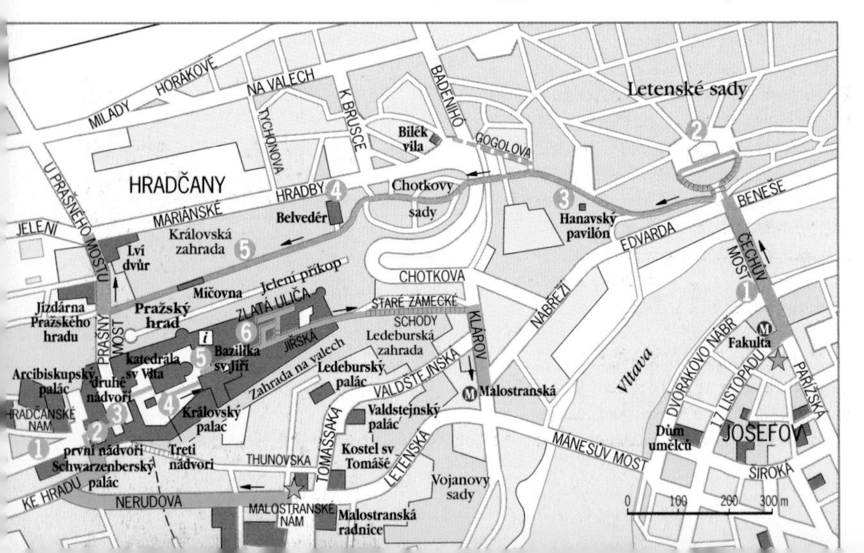

intended to symbolise the return to democracy. In the hillside beneath your feet is a nuclear bunker – the least deserving members of the population planned to retire here in the event of Armageddon. It was actually used to house the city's potato supply. The view of the Vltava and Staré Město is stunning.

3 HANAVSKÝ PAVILÓN

A short walk south brings you to the extravagantly ornate Hanavský pavilón (Pavilion), originally erected for the Paris World Exhibition of 1878, and re-erected here in 1898. It has a charming little café on the terrace.

Walk on south through leafy Letná until you come to the footbridge over Badeniho, beyond which the Belvedér (Belvedere Palace) can be seen, shrouded in trees at the end of an alley. A detour to the right before reaching the footbridge down Gogolova and across the junction brings you to the Bílek villa at Mickiewiczova 1, former home and studio of František Bílek. His interesting symbolist sculptures and the furniture he designed himself are on display.

4 BELVEDERE

The Královský letohrádek (Summer Palace) or Belvedere (see page 32) was built on the orders of Ferdinand I as a gift to his wife, Queen Anne. Its terrace was used by Rudolf II's Danish astronomer, Tycho Brahe. In the garden on the west side of this lovely Renaissance building is the bronze 'singing fountain', so-called after the sound of its gently splashing water.

5 KRÁLOVSKÁ ZAHRADA

The Royal Gardens beyond (open: May to September, Tuesday to Sunday 10am–6pm, admission charge) form one of the best kept parks of Prague. Experimental modern sculptures (said to have been personally chosen by President Havel) currently mingle with formal parterres, manicured lawns and noble trees.

Halfway up on the left is the Míčovna (Ball Game Court) with sgraffito decoration on the exterior walls. Parallel to the gardens to the southeast is the Jelení příkop (former 'Deer Moat'), where the kings used to corral their red deer. At the southwest end is the Lví dvůr (Lion's Court). This was once Rudolf II's menagerie, where lions and tigers were kept in cages that had to be provided with luxurious and expensive heating against the bitter Prague winters.

You come out of the gardens at the west end, opposite the baroque Jízdárna Pražského hradu (former Riding School) that is now used for exhibitions. Leaving the Prašny most ('Powder Bridge') that leads into the castle complex on your left, turn right along U Prašného mostu for Mariánské hradby and the no 22 tram.

Hanavský Pavilion

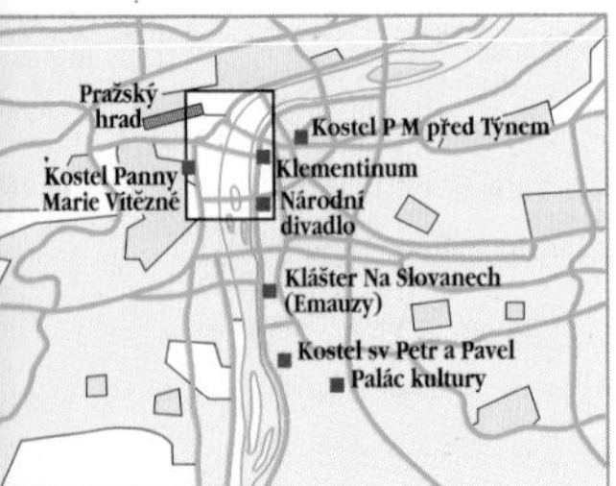

Malá Strana

This walk covers many of the most interesting sights of the Malá Strana (Lesser Quarter). The nobles built their great palaces here to be close to (or, in Wallenstein's case, to rival) the Royal Court of Hradčany. *Allow 2 hours.*

Begin at the Malostranská metro station (also the tram stop for nos 12 and 22), and turn left into Valdštejnská.

1 VALDŠTEJNSKÁ

The street is filled with the palatial residences of the German nobility imported by the Habsburgs. The Valdstejnsky palác (Wallenstein Palace) occupies the entire east side of Valdštejnská náměstí at the end of the street. Albrecht von Waldstein (Wallenstein to the English) was the greatest imperial commander in the Thirty Years' War. His vaunting ambition proved his downfall: the Emperor Ferdinand had him assassinated when he set his sights on becoming king of Bohemia.

From Valdštejnské náměstí continue into Tomášská and Malostranské náměstí (see pages 78–9).

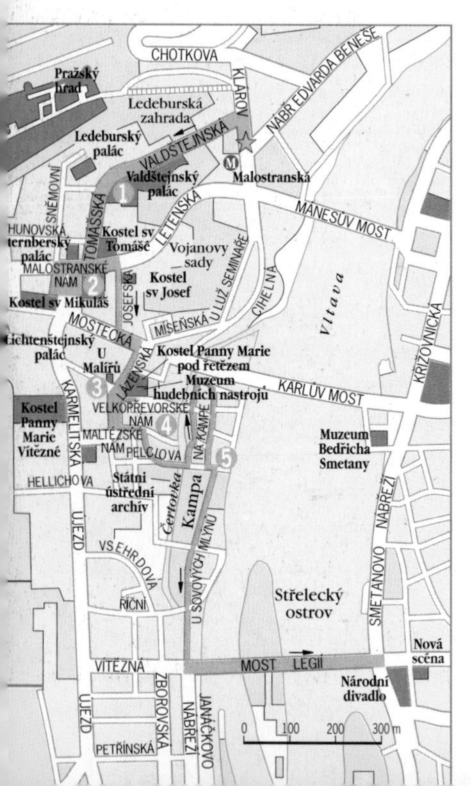

2 MALOSTRANSKÉ NÁMĚSTÍ

Malostranské náměstí (Lesser Quarter Square) manages to be both noble and intimate. The whole of the west side is occupied by the Lichtenstejnský palác (Liechtenstein Palace), home of the Liechtenstein who condemned the Protestant leaders to death in 1621, while in the Smiřický Palace to the north was hatched the plot that led to the defenestration of Ferdinand's Catholic councillors in 1618. Louring over these nests of intrigue is baroque Kostel svatého Mikuláše (St Nicholas Church – see page 79), the greatest masterwork in Prague of the Dientzenhofers, father and son.

Walk east along Letenská, turning

immediately right into Josefská and passing the Kostel svatého Josef (St Joseph Church) with its narrow Dutch-type façade. Turn left into Mostecká (always awash with tourists heading for the Karlův most – Charles Bridge) and then right into Lázeňská.

3 LÁZEŇSKÁ

A few minutes on foot brings you to the area associated with the Maltese Knights (of the Order of St John), as recalled in the names of its two diminutive and picturesque squares: Maltézské náměstí (Maltese Square) and Velkopřevorské náměstí (Grand Prior's Square). A relic of their function as keepers of the first bridge across the Vltava is the chain hung above the high altar of their church, Kostel Panny Marie pod řetězem (Church of Our Lady below the Chain, accessible for Sunday mass at 10.30am, which is held in French). At no 11 Maltese Square is the elegant little U Malíru (Painters' Tavern), now an expensive French restaurant.

Continue to the adjoining Velkopřevorské náměstí (Grand Prior Square).

4 VELKOPŘEVORSKÉ NÁMĚSTÍ

The Palace of the Grand Prior used to host the Museum Hudebních nastrojů (Musical Instruments Museum). A back wall has become an unofficial shrine to John Lennon, with portraits of the philosopher Beatle, making him look a bit like a Byzantine saint.

Retrace your steps to Maltézské náměstí and enter Nosticova. Turn left on Pelclova and cross the bridge to Kampa Island.

5 KAMPA ISLAND

The bridge spans the Čertovka, or Devil's Stream, named after the tricky sprite thought to inhabit it (or possibly just an abusive reference to a famously grumpy local washerwoman – opinions differ).

It is worth taking a turn round the delightful 'island' and perhaps stopping for a drink at one of the pavement cafés below Charles Bridge on Na Kampe, where there is also a shop selling attractive pottery.

Leave Kampa by the southern end and climb up on to most Legií (the Bridge of the Legions). On the far side are the Národní divadlo (National Theatre) and tram stops.

The serene beauty of Kampa Island as seen from the Charles Bridge

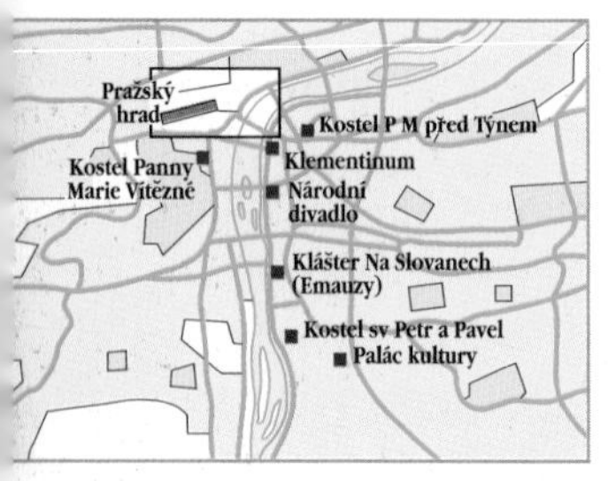

A Stroll Through Castle Hill

Pražský hrad (Prague Castle), was the spiritual and political focus of the city for most of its history. There has been a fortress and a church here since the 9th century, and from the 12th century until the Republic it was officially the seat of the kings of Bohemia. A detailed description of the buildings is on pages 46–57. See map on page 126 for route. *For a stroll, allow about an hour; comprehensive sightseeing requires at least half a day.*

Approach Hradčany up Nerudova and Ke Hradu from Malostranské náměstí (trams 12 and 22).

1 HRADČANSKÉ NÁMĚSTÍ

Hradčany Square (see page 132) is flanked by noble buildings such as the Arcibiskupský palác (Archbishop's Palace) on the north side. Its late incumbent, the heroic Cardinal Tomášek, lived just long enough to see the collapse of the godless regime he withstood. The scion of the former owners of the impressive Schwarzenberský palác (Schwarzenberg Palace) on the south side, Karl Schwarzenberg, has also enjoyed a remarkable rehabilitation as personal adviser to President Havel. Havel's influence may be seen in the uniforms of the guards on duty

below the statues of battling giants atop the first gateway. The costume designer for Miloš Forman's *Amadeus* was commissioned by the President to find a replacement for the dreary tunics worn formerly.

2 PRVNÍ NÁDVOŘÍ

The two huge flagstaffs in the První nádvoří (First Courtyard) are made from fir trees and date from the remodelling of the castle area by the Slovene architect, Josip Plečnik, during the First Republic. Ahead you can see the plain baroque façade of the Presidential Apartments (to the south) and (to the north) the Spanish Hall and Castle Gallery. All this, together with the east wing of the next courtyard, is the work of Maria Theresa's architect, Nicolo Pacassi.
Pass through the Matthias Gate.

3 DRUHÉ NÁDVOŘÍ

The Matyasova brána (Matthias Gate), leading to the Druhé nádvoří (Second Courtyard), is a triumphal arch named after the brother of Rudolf II. The neo-classical building on the right is the Kaple svatého Křízě (Chapel of the Holy Cross), once the treasury, now a gallery.

4 TRETÍ NÁDVOŘÍ

The Tretí nádvoří (Third Courtyard), beyond, is dominated by the Katedrála svatého Víta (Cathedral of St Vitus). Walk across the space to get a better view of the South Tower and the early Gothic architecture of the eastern end of this great edifice constructed by Petr Parléř and his sons. If you walk on through the courtyard, the Královský palác (Old Royal Palace), with its celebrated Vladislavský sál (Vladislav Hall) and

Left: statue of St George, Prague Castle

Visitors are drawn to the imposing Matthias Gate entrance to Prague castle

Riders' Staircase is ahead of you. On the right, Plečnik's green cylindrical canopy entices you down some steps to the Zahrada na Valech (Rampart Garden).
Enter Jiřské náměstí.

5 BAZILIKA SVATÉHO JIŘÍ

The last courtyard – St George's Square – contains the Basilica of St George, one of the finest surviving Romanesque buildings in Central Europe. The origins of this and the adjacent monastery lie in the 10th century. The monastery houses the National Gallery's collection of early Bohemian art, and the basilica can be visited or enjoyed at leisure if you attend one of the regular chamber music concerts held here.
Now take Jiřská, beside the church, and turn left as soon as you can to reach the famous Zlatá ulica ('Golden Lane').

6 ZLATÁ ULICA

In the 19th century, the tiny 16th-century houses that line the street were supposed to have been the dwellings of Rudolf II's alchemists. Franz Kafka lived here for a while.
Return to Jiřská and descend the steps at the northwestern tip of Castle Hill, which lead eventually to Klárov, the Malostranská metro station and no 22 tram.

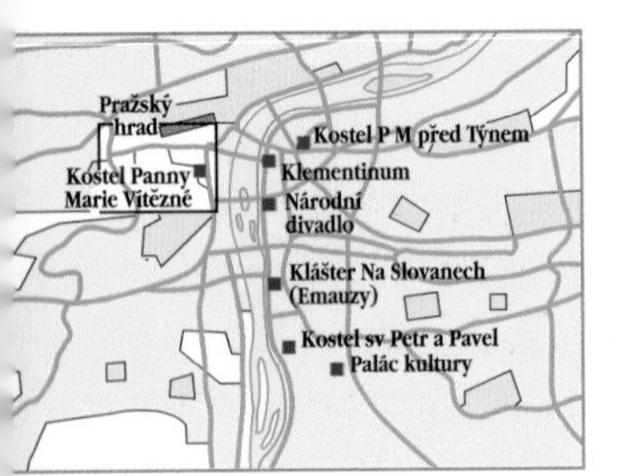

Hradčany Hinterland

The tranquil backwater of the Castle District seems frozen in time. The bustling medieval town fell victim to the great fire of 1541, but in the 17th century the Catholic beneficiaries of the Counter-Reformation built grandiose palaces which still abound. *Allow 2½ hours.*

The walk begins in Hradčanské náměstí, reached via Nerudova and Ke Hradu from Malostranské náměstí (trams 12 and 22).

1 HRADČANSKÉ NÁMĚSTÍ

Apart from the Arcibiskupský palác (Archbishop's Palace) and the Schwarzenberský palác (Schwarzenberg Palace – see page 130), the western end of the square is dominated by the Toskánský palác (Tuscany Palace). North of it, on the corner of Kanovnická, is the more modest Martinický palác (Martinic

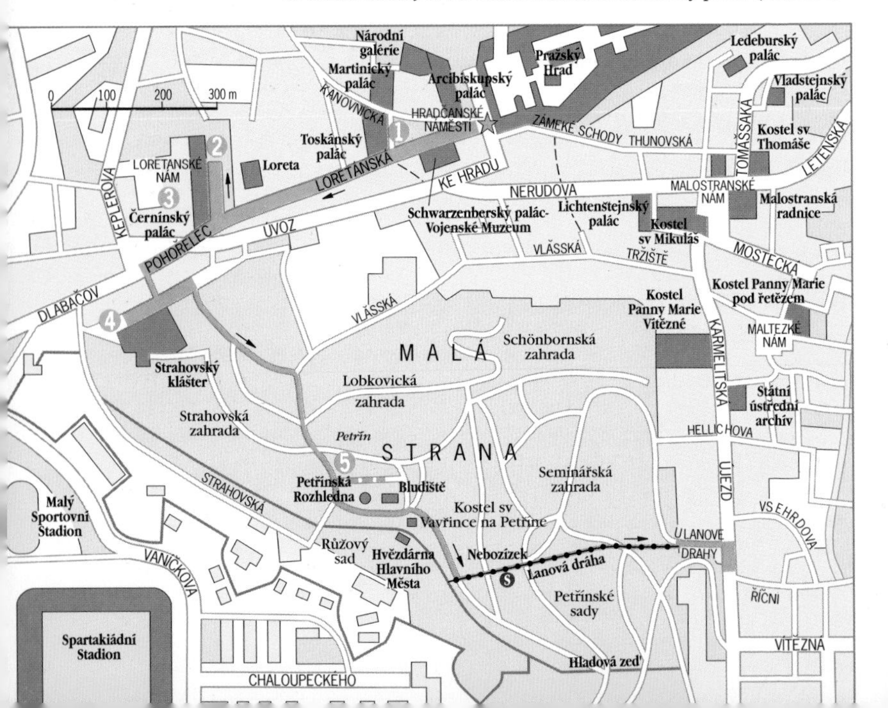

Palace), notable for its lively sgraffito decoration. In the centre of the square is a baroque Marian Column, erected in thanksgiving for deliverance from the plague. Worth a look, also, is the heavily ornate wrought-iron street lamp, one of two in Prague.

Walk west along Loretánská and turn right down to the shrine of Loreta.

2 THE LORETA

The Loreta is inspired by the famous Italian original (Loreto) that claims to possess the house in Nazareth where the Annunciation took place. The most entertaining part is the cloisters, lined with user-friendly saints whose responsibilities range from the alleviation of gallstones (St Liborius) to the return of lost property (St Anthony). The Santa Casa contains pictures of scenes from the life of the Virgin, and claims to possess beams and a brick from the real Loreto. The Church of the Nativity is notable for the depiction of St Agatha carrying her severed breasts on a plate and the treasury for a fabulous diamond monstrance designed by Fischer von Erlach.

Walk back up to the junction with Pohořelec and turn right.

3 ČERNÍNSKÝ PALÁC

On the corner is the vast rusticated façade of the Cernín Palace, the largest in Prague. During the First Republic it was taken over by the Foreign Ministry and it was from here that Jan Masaryk, the only non-Communist left in the cabinet of Klement Gottwald, plunged suspiciously to his death on 10 March 1948.

Continue down Pohořelec, passing the junction with Úvoz, then turn left into a cobbled square with limes and acacias, around which are grouped the buildings of the Strahovský klášter (Strahov Monastery – see page 112).

4 STRAHOVSKÝ KLÁŠTER

Since the revolution of 1989 the Premonstratensian monks have returned to their ancient home, which was founded in 1140. The high point of any visit is the frescoed library, particularly the Philosophical Hall with its painted ceiling ('The History of Humanity') by Franz Anton Maulpertsch.

Leave the monastery by an archway in the eastern wall and walk south through the Strahovská zahrada (Strahov Gardens). The Vltava and Malá Strana unfold before you. In 5 to 10 minutes you come upon steps to your right, which lead up to Petřin, (see pages 102–3).

5 PETŘÍN

This is an especially pleasant part of the walk through pear and plum orchards. At the top of the crumbling steps is Prague's mini-Eiffel Tower, worth climbing for the view. Near by are the Bludiště (Mirror Maze) and the Observatory. From the summit of the hill you can see a section of the so-called Hladová zed ('Hunger Wall') to the south. It is said to have been built as a job-creation scheme by Charles IV between 1360 and 1362; others point out that it was paid for by the expropriation of the Jews.

The funicular railway (*lanová dráha*) terminus on the summit is approached through a rose garden. The railway is now electric, but until the 1960s worked by water pressure. Halfway down you can alight at Nebozízek Station, where there is a restaurant with another good view.

Descend to Újezd, where you can pick up trams to the city centre.

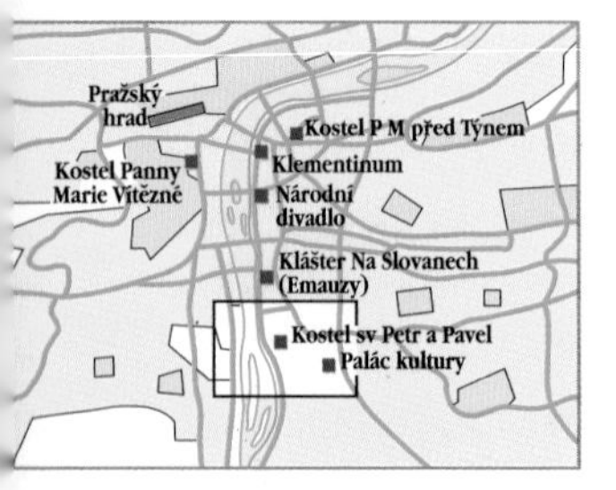

Vyšehrad

This walk takes you through the ancient citadel of Vyšehrad on its great rock, and offers Prague's finest views of the Vltava. ***Allow 2 hours.***

Take the metro (line C) to Vyšehrad. Walk west along the terrace of the ugly modern Palác Kultury (Palace of Culture), then enter Na Bučance.

1 TABOR BRÁNA

The Tabor Gate of Vyšehrad (meaning 'High Castle') is at the end of the road. Legend says that the Slavic tribes originally settled on this windy outcrop when they reached the Vltava. It was here that Libuše, daughter of an early chieftain, is supposed to have had a vision prophesying the foundation of Prague. She is also said to have had numerous discarded lovers hurled from the top of the cliff in to the Vltava. However, the archaeological evidence suggests that Hradčany was settled before Vyšehrad.

Continue along V pevnosti.

2 LEOPOLDOVÁ BRÁNA

You pass through the Leopoldová brána (Leopold Gate),

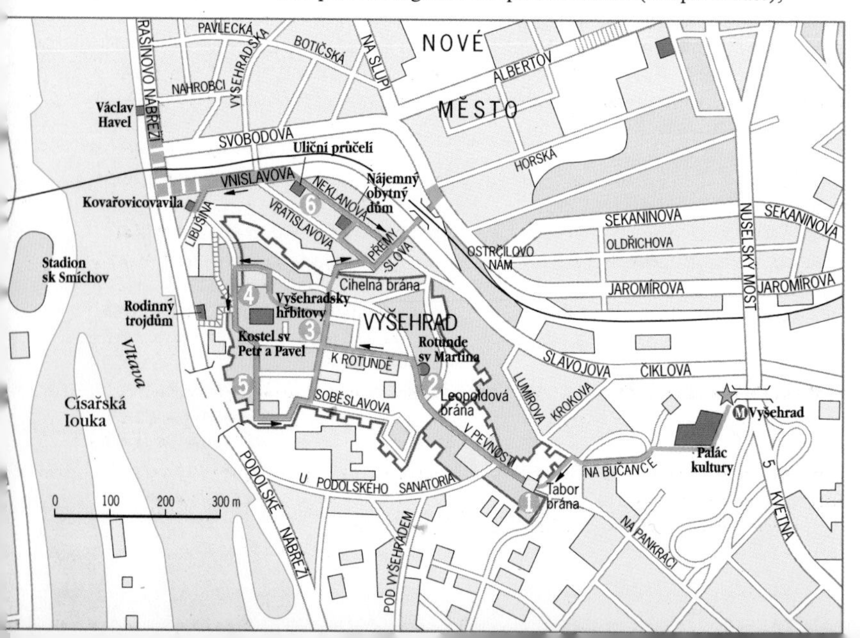

beyond which, on the right, is the earliest and best preserved of Prague's Romanesque rotundas (Rotunda svatého Martina/St Martin's). It probably served as a cemetery chapel and was built some time in the 11th century.
Bear left into K rotundě and then right down to the east gate of the Vyšehradsky hřbitov (Vyšehrad cemetery – see page 34, opening times vary but are usually between 8am and 6pm).

3 VYŠEHRADSKÝ HŘBITOVY

Leading intellectual figures of the Czech national revival in the 19th century – musicians, writers, sculptors, painters and scientists – are buried here, but soldiers and politicians are excluded. The most impressive part is the Slavín pantheon commemorating leading figures in the arts – a plinth topped by a sarcophagus guarded by a winged genius.

4 KOSTEL SVATÉHO PETR A PAVEL

A walk through the cemetery. brings you to the front of the Kostel svatého Petr a Pavel (Church of St Peter and St Paul), which has undergone alteration in every conceivable style since it was built in the 11th century. A baroque gate to the south leads into a park, presided over by Josef Myslbeck's gigantic patriotic sculptures, originally made for the Palackého most (Palacký Bridge). Přemysl and Libuše, founders of the Bohemian dynasty, are nearest to you on the left.

5 WESTERN BASTIONS

From here a circuit of the western bastions can be made, offering marvellous views of the Vltava streaming sluggishly far below. Note the ruins of a watch-tower precariously perched on the almost sheer rock, romantically called 'Libuše's Baths', but actually where goods were hauled up from the river in the Middle Ages.
Cut back across the park to St Peter and St Paul and continue down to the Cihelná brána (north gate) of the citadel. Those who prefer to head for home at this point can do so by following the Přemyslova down towards Na slupi). Otherwise descend towards the town via Vratislavova, then turn right into Hostivítova.

6 NÁJEMNÝ OBYTNÝ DŮM

At the bottom, on the corner of Hostivitová and Neklanova (no 30), is Nájemný obytný dům, a celebrated Cubist building designed by Josef Chochol (see pages 40–1). Other such buildings can be seen at Neklanova 2, and by walking west along Vnislavova to Libušina no 3. The latter is Chochol's ambitious Kovařovičova Villa, with a remarkable façade conceived in diamond shapes and curious zig-zag railings round the garden at the rear.

Re-enter Vnislavova and head towards the embankment. Turn right into Rašínovo nábřeží, formerly the Engels Embankment, and a short way along at no 78 is where Václav Havel lived. The art nouveau apartment block had been built and owned by his grandfather, a successful contractor at the turn of the century and has now been returned to the family. The house has become a place of pilgrimage for Havel fans, who have decorated the walls with daubed eulogies.
Retrace your steps along Vnislavova and Nekanova, turning left into Přemyslova. Cross the busy Vnislavova and pass under a crumbling railway bridge, beyond which is Na slupi. Trams for the centre can be boarded here.

Excursions

ČESKÝ ŠTERNBERK (Bohemian Sternberg)

Southeast of Prague is the impressive stronghold of Český Šternberk, built in 1242 on a promontory overlooking the point where the Sázava is joined by its Blanice tributary.

The interior of the castle may be visited with a guided tour and is notable for the baroque stucco made by Italian craftsmen between 1660 and 1670. Highlights of a visit are the engravings on the staircase of the Thirty Years' War, the Yellow Chamber with Carlo Brentano's stucco and a collection of silver miniatures.

The castle lies 45km southeast of Prague off motorway D1. Open: April to October, Tuesday to Sunday 9am–4pm, but the new administration is still making changes to opening times. To check, tel: 0303 55101. Closed: Monday. Admission charge.

HRAD KARLŠTEJN (Karlštejn Castle)

The most celebrated Gothic castle of Bohemia was founded by Emperor Charles IV in 1348 and completed by 1357. It was planned as an Imperial and Christian sanctuary. On the highest of its three levels was the Chapel of the Holy Rood, which was also a treasury, housing the Crown Jewels of the Holy Roman Empire (now in Vienna) and the Bohemian Regalia. The castle's architect

Gothic Karlštejn Castle dominates the surrounding countryside

REGIONAL MAP

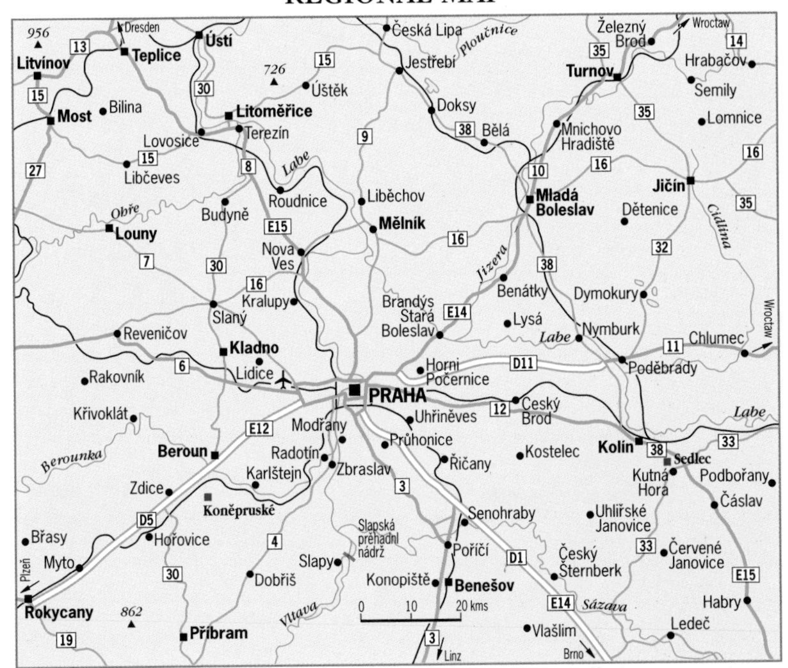

is thought to have been Matthew of Arras, who began the building of St Vitus Cathedral.

You approach the castle on foot from the car and coach park 2km below. The interior must be visited with a guided tour: highlights include the **Audience Hall,** with handsome wooden panelling and a coffered ceiling, and the **Luxembourg Hall**, which contains a model of how the room looked before the 16th century. The **Church of the Virgin Mary** on the second floor of the north tower is notable for the cycle of frescos showing the Emperor receiving relics from various distinguished donors, and another cycle depicting the Apocalypse. Adjoining the church is **St Catherine's Chapel**, with decoration recalling that of the St Wenceslas Chapel in St Vitus Cathedral.

The **Kaple svatého křízě** (Chapel of the Holy Rood) should be the high point of the tour, if it has been reopened after restoration. Its 6m-thick walls are encrusted with 2,200 semi-precious stones and lined by 128 wooden panels painted by the court painter, Master Theodoric.

Trains leave for Karlštejn from Smíchovské nádřaží in Prague every hour and the journey takes about 45 minutes. By road it is 28km to the southwest. Tel: 0311–94211. Castle open: March, April and October to December, 9am–noon, 1–4pm; May and September, 8am–noon, 1–6pm. Admission charge. Closed: January and February.

ZÁMEK KONOPIŠTĚ
(Konopiště Castle)

This magnificent but gloomy Gothic and Renaissance castle was acquired in 1889 by Archduke Franz Ferdinand d'Este, the heir to the imperial throne. In 1907 he employed an English botanist to cross-breed roses at Konopiště in order to produce a black variety, a commission that provoked Delphic warnings about black roses bringing war and death. It apparently took the botanist until 1914 to cultivate the rose. That same year Franz Ferdinand was assassinated in Sarajevo and World War I began.

The car park is 2km below the castle. You enter through the east tower and walk through a baroque gateway by F M Kaňka with statues by Matthias Braun. The impressive moat is now perambulated by languid peacocks, but in Franz Ferdinand's day it was occupied by bears.

The guided tours of the rambling interior can be wearisome. Highlights include the St George's Museum, some fine furniture, Habsburg *memorabilia* and the royal bathroom. The weapons collection is one of the biggest in Europe with a number of historically important items.

The 90-hectare park is well worth visiting and has an attractive rose garden, a deer park and a lake. Weather-beaten baroque statues help to create a romantic atmosphere.

Konopiště is 45km south of Prague. A train runs to nearby Benešov (2km) from Praha hlavní nádraží or a bus runs from Florenc terminal. Tel: 0301–21366. Open: April, September and October, Tuesday to Friday 9am–3pm, weekends to 4pm. May to August, Tuesday to Friday 9am–4pm, weekends to 5pm. Admission charge.

Mělník Castle overlooks the River Labe (Elbe)

ZÁMEK KŘIVOKLÁT
(Křivoklát Castle)

Ancient Křivoklát is perched on a ledge jutting out of the forests above a tributary of the Berounka river. The castle was first mentioned in records of the year 1110, and became a Přemyslid residence from the reign of Otakar II (1252–78). Much of what is now to be seen is the result of late Gothic reconstruction under the Jagiello king, Vladislav II (see page 56). Rudolf II's English alchemist, Edward Kelley, imprisoned here, may have died leaping from a tower window in an attempt to escape.

The castle tour takes about an hour and is well worth it for the remarkable Gothic architecture, the paintings and sculptures, the chapel and the library.

40km west of Prague. Buses run regularly from Praha-Dejvice and take about 1½ hours. Trains run from Smíchovské nádraží to Beroun, where you must change for Rakovník. Tel: 0313–98120. Open: April, September and October, Tuesday to Sunday 9am–4pm; May to August, 9am–6pm. Admission charge.

Hunting is the theme of Konopiště Castle

MĚLNÍK

The wine-producing town of Mělník has a delightful position overlooking the confluence of the Vltava and Labe (Elbe) rivers. The surrounding vineyards were first planted by Charles IV, who brought French wine expertise to the region.

Sights in the town include the Gothic church, the former Lobkowitz Castle, which houses a collection of baroque paintings, and a Regional Museum. Ideally you should time a visit to include a meal, which can be enjoyed on the terraces overlooking the rivers; this would also provide an excuse to sample the local wine, of which the Tramín is particularly good.

32km north of Prague. The bus service from Praha-Dejvice takes about one hour. Tel: 0206–2421 (Lobkowitz Castle). Castle and museum open: March, April and September, Tuesday to Sunday 9am–4pm; May to August, 8am–noon, 1–5pm. Admission charge. Closed: October to February.

KUTNÁ HORA (Kuttenberg)

The name of Bohemia's one-time second city means 'mining mountain', a reference to the deposits of silver and copper ore on which its prosperity was founded. The mines were vigorously exploited from the late 13th century, bringing a rapid increase in wealth and attracting miners (chiefly Germans) from outside Bohemia. The royal mint was founded here at the beginning of the 14th century and King Wenceslas II summoned experts from Florence to advise him on his coinage. They produced the *pražské groše* (Prague Groschen), a silver coin regarded as sound currency all over Central Europe for several centuries.

Church statuary, Kutná Hora

The prosperity of Kutná Hora – evident from its ambitious cathedral and handsome burghers' houses – lasted until the mid-16th century. Dwindling reserves and flooding put paid to the mining and the town rapidly fell into decline.

Chrám Svaté Barbory (Cathedral Of St Barbara)

This great Gothic cathedral was financed by the miners and dedicated to their patron saint. Building, initiated by Petr Parléř in the late 14th century, was halted by the Hussite wars. At the end of the 15th century, two of the greatest architects of Prague, Matthias Rejsek and Benedikt Ried, produced the marvellous late Gothic vaulting inside. There are frescos showing work in the mint (in the Chapel of the Mint-Men), and toiling miners (in the ambulatory).
Open: Tuesday to Sunday 8am–noon, 1–5 pm in summer; 9am–noon, 2–4pm in winter. Admission charge.

Mining Museum

A walk down sculpture-lined Barborská ulice brings you to the Silver Mining Museum. Its display rooms are in the Renaissance house of one John Smíšek, who grew rich by exploiting a private (and illegal) mine in the 15th century. There is an exhibition about the history of mining here, but most visitors head for the medieval mine down the lane behind the building. At the entrance to the 250m of tunnels is a *trejv*, or horse-drawn hoist.
Barborská ulice 28. Open: April to September, Tuesday to Sunday 8am–noon, 1–5pm; October till 4pm. Admission charge.

The Cathedral of St Barbara, the patron saint of miners

Vlašský Dvůr (The Italian Court)
Wenceslas II's mint – known as the Italian Court, after the king's Florentine advisers – later became a royal residence. The bricked-up outlines of the mintmen's workshops can still be seen, together with the chapel and other parts of the former palace. Across the square to the west is the Chrám svatého Jakuba (Cathedral of St James), a Gothic structure built in 1420 with a baroque interior.
Havlíčkovo náměstí. Open: daily 8am–5pm in summer, 8am–4pm in winter. Admission charge.

Other sights worth a visit include Matthias Rejsek's magnificent Stone Fountain (1495) in Husova ulice and the Kamenný dům (Stone House) off Hornická ulice, with its richly decorated front. The latter is a Gothic burgher's house whose rooms can be visited.

SEDLEC
Three kilometres to the north of Kutná Hora is Sedlec, where the star attraction is the Cistercian *kostnice* (ossuary). In the last century František Rint used its 40,000 bones to create one of Europe's most macabre spectacles – an interior decorated with a 'chandelier', 'bells', 'urns', even a Schwarzenberg 'coat of arms', *all* made from human bones.
Buses 1 and 4 run from Kutná Hora to Sedlec. Ossuary open: Tuesday to Sunday 8am–noon, 1–5pm in summer, 9am–noon, 1–4pm in winter.

Kutná Hora is 65km east of Prague. Buses run from the Želivského or Florenc terminals in Prague and take about 1½ hours. Trains run sparingly from Masarykovo or Hlavní nádraží, but the town's main station is far from the sights.

LIDICE

The village of Lidice was razed to the ground by the Nazis on 10 June 1942 in revenge for the assassination of the Bohemian governor, Richard Heydrich, by the Czech resistance. There is a small Memorial Museum and a Rose Garden of Friendship and Peace planted in 1955.
22km northwest of Prague. A regular bus service leaves from Praha-Dejvice. Museum open: April to September, daily 8am–5pm; October to March, 8am–4pm.

TEREZÍN

Northwest of Prague in the former Sudetenland, Joseph II built a fortified town – in practice a huge barracks and prison – in 1780 and named it 'Theresienstadt'after his mother, Maria Theresa. In June 1942 the inhabitants were driven out by the Nazis and the main part of it was turned into a ghetto for Jews.

The Nazis used Terezín for their propaganda, allowing the doomed Jews to pursue cultural activities such as music and drama and building gleaming facilities to show visitors from the Red Cross. The Ghetto Museum has an informative (and harrowing) display, including clips from the Nazi propaganda film *Hitler gives the Jews a Town.*

On the other side of the river that divides the town the Malá Penost (Lesser Fortress) may also be visited. A prison under the Habsburgs (Gavrilo Princip, the murderer of Archduke Franz Ferdinand at Sarajevo in 1914 languished here), it was later used as a concentration camp and extermination centre. There is an exhibition in the former house of the camp commandant. A short documentary film is shown and you can tour the cells.
60km northwest of Prague and 1 1/2 hours by bus from Florenc bus terminal. Weekly tours also leave from Josefov – inquire at the ticket office of the State Jewish Museum. Ghetto Museum open: daily 9am–6pm, the Malá Pevnost daily 9am–5pm.

Sombre monument to the Holocaust, Terezín

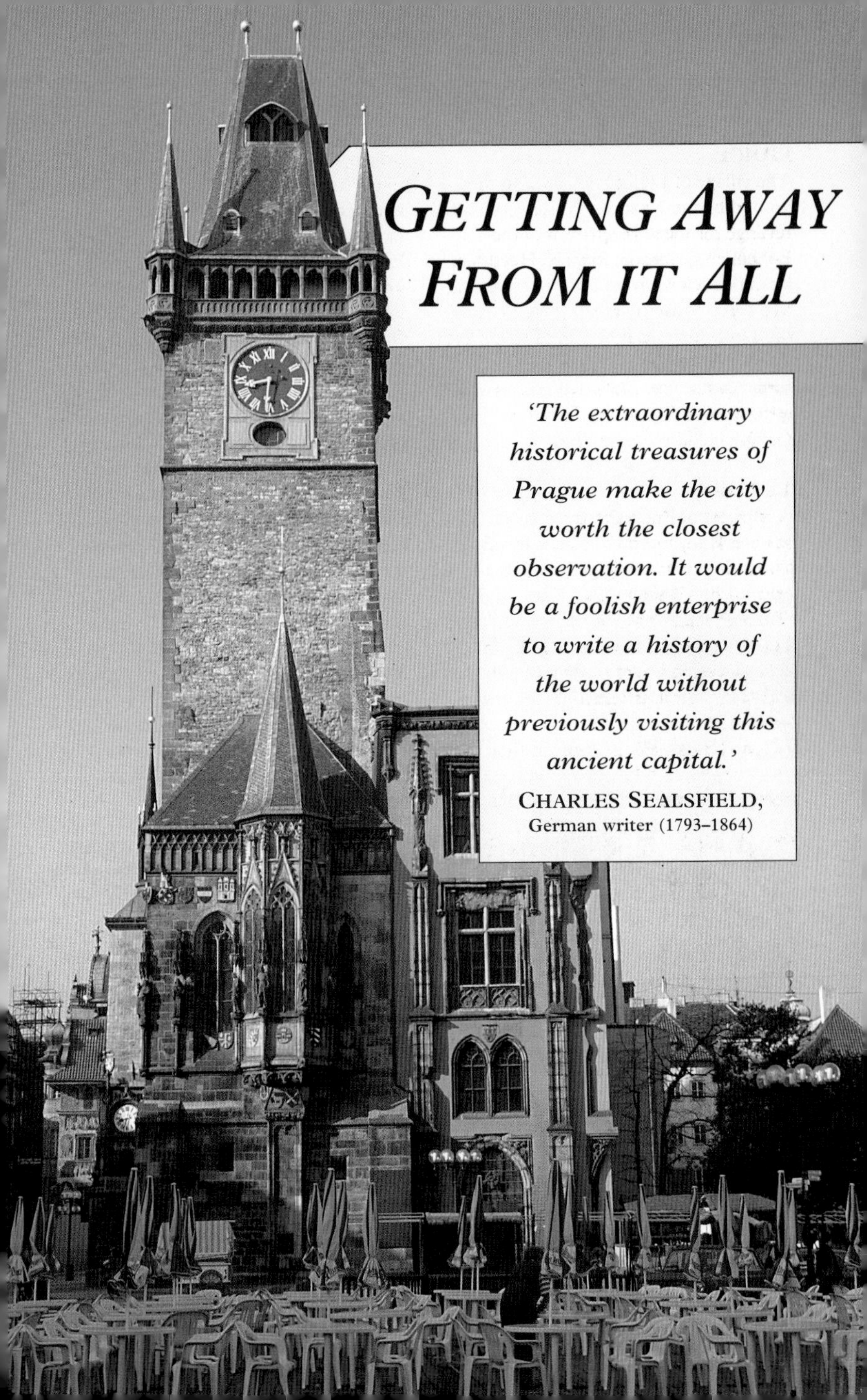

Getting Away From it All

'The extraordinary historical treasures of Prague make the city worth the closest observation. It would be a foolish enterprise to write a history of the world without previously visiting this ancient capital.'

CHARLES SEALSFIELD,
German writer (1793–1864)

Getting Away From it All

BOAT TRIPS ON THE VLTAVA

From May to September (as long as there is enough water) excursions run both north and south on the Vltava. To the north they go as far as Roztoky (1 hour 20 minutes) and to the south as far as Štěchovice (3 hours) and Slapy Dam (4 hours). Shorter trips offer a 'panorama of Prague' (50 minutes to 2 hours).

A favourite destination for Praguers who want to get away from it all is **Slapská přehradní nádrž** (Slapy Dam), reached via a picturesque 33km stretch of the Vltava valley. The 65m-high dam was built in 1954 and has 40km of reservoir stretching behind it, a paradise for water sports and fishing.

BÍLÁ HORA (White Mountain)

On the western outskirts of Prague is White Mountain, where the most decisive battle of Bohemian history was fought on 8 November 1620. On this limestone plateau the Protestant mercenaries of Bohemia under Count Thun were catastrophically defeated by an imperial Catholic army under Maximilian of Bavaria. The elected Bohemian king (Frederick of the Palatinate) fled, opening the way to three centuries of Habsburg rule. Czechs see this date as marking the extinction of their independence – not to be revived until 1918 – and the beginning of the *temno* (darkness).

A small monument marks the site of the battle. Near by, to the south, is the **Chrám Panny Marie Vítězné** (Church of St Mary the Victorious), built as a chapel in 1622 and rebuilt as a pilgrimage church between 1704 and 1714. Inside are fine baroque frescos. *Trams 8 and 22 to end stop at Řepy. The church is next to the tram stop; the monument is further to the west and along Nad višňovkou. This trip can easily be combined with a visit to the Star Castle (see page 67).*

KONĚPRUSKÉ JESKYNĚ – ČESKÝ KRAS (Koněpruské caves and the Bohemian karst)

The Bohemian karst is a protected ecological area rich in rare flora, but most visitors go for the stalactite and stalagmite limestone caves. Since their discovery in 1950, 800m of labyrinthine chambers have been made accessible to the public. A further exciting discovery was the remains of an illegal mint dating to the second half of the 15th century; (there is a small exhibition about it). *Open: April to October, Tuesday to Sunday. A visit could be combined with an excursion to Karlštejn; from Srbsko (one stop beyond Karlštejn on the railway) a yellow-marked path leads to the caves (about 3km). A bus connects them with the station at Beroun, 50 minutes by train from Praha Hlavní nádraží.*

KUNRATICKÝ LES (Kunratice Woods)

This extensive area of pleasant woodland southeast of the city is criss-crossed with asphalted paths. A moufflon herd is said to run wild here; there is a mini-zoo at the forester's hut with a few wild boar, deer and pheasants.

You can walk to the ruin of Nový hrad, known as the 'Stone of Wenceslas'

BOAT TRIPS
Excursion boats are run by Pražská Paroplavební Společnost, (PPS), and leave from the mooring at Palackého most (tel: 29 38 03/29 83 09). The nearest metro station is Karlovo náměstí.

because Wenceslas IV built it as a hunting lodge. Supposedly it was here, in 1419, that the king had a fatal stroke on hearing of the defenestration of Catholic councillors. There is a restaurant named after the choleric king, U krále Václava IV.

On the west side of the woodland flows the Kunratice river. Several of its pools make informal bathing beaches, mostly around Šeberák. Near here the first nude beach was instituted when such things were still considered extremely daring.
Metro (line C) to Roztyly.

Monument to the Battle of the White Mountain

PRŮHONICE

Most visitors to the 19th-century palace of Průhonice, on the southeast periphery of Prague, are drawn by the marvellous landscaped garden. It contains 700 different plants and shrubs, including many alpine species and rhododendrons.
A regular ČSAD bus service to Průhonice runs from Praha-Chodov. By car, turn off the D1 motorway to the village, about 10km after leaving the city.

STROMOVKA/VÝSTAVIŠTĚ

The Holešovice district of Prague has an extensive exhibition area (Výstaviště) for which remarkable buildings were designed in 1891 (see page 29). Apart from the regular fairs and exhibitions held here the place is known for its funfair (*Dětský svět*), its Planetarium (tel: 37 17 46 – generally open 8am–6pm, but hours vary) and the Maroldovo panoráma (open daily 9am–5pm), a diorama of the Battle of Lipany (1434).

Stromovka Park, stretching to the west of Výstaviště, was planted on the orders of Rudolf II in 1593 and an artificial lake was created.
Metro (line C) to nádraží Holešovice, or trams 5, 12 and 17 to Výstaviště. The latter is open daily from 2pm, at weekends from 10am.

Relaxing in Stromovka Park

Worth a visit, the landscaped garden at Průhonice

VLTAVA ISLANDS

Eight islands remain from twice that number before regulation of the Vltava in the late 19th century. The most picturesque is **Kampa**, an island only by virtue of the millstream that separates it from the Malá Strana bank.

Opposite Vyšehrad are **Císařská louka** (the Emperor's Meadow) on the Smíchov side (a venue for water sports) and the **Veslařský ostrov** off the Vyšehrad/Podoli shore (the haunt of sailors and skullers). The **Slovanský ostrov Žofín** (between Jiráskův most and Most Legií) has a station for hiring out rowing boats in summer. All the islands are accessible on foot.

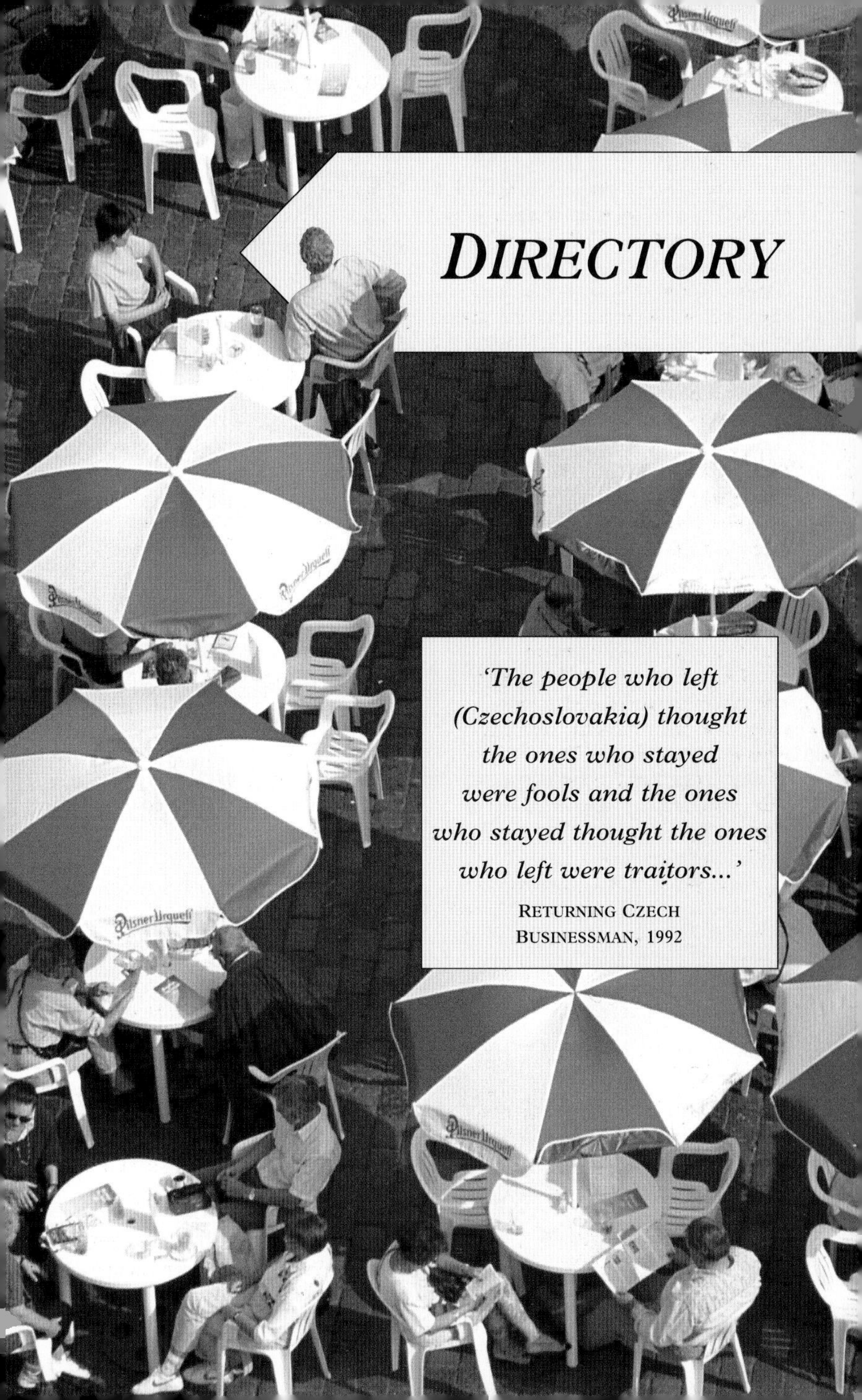

DIRECTORY

> *'The people who left (Czechoslovakia) thought the ones who stayed were fools and the ones who stayed thought the ones who left were traitors...'*
>
> RETURNING CZECH BUSINESSMAN, 1992

Shopping

Prague's retail trade is in a state of flux. The future of the state-owned chains must be in doubt as private competition enters the market. Service has improved, but prices have increased enormously: some have gone up 300 per cent in three years.

A good example of transformation for the better is the Bata shoe emporium (Václavské náměstí 6, tel: 2421 8133) which offers a wide range of footwear at reasonable prices. This store accepts credit cards, but generally cash is king.

What to buy

There are a number of items for which Bohemia is famous and which make good souvenirs, particularly glassware.

Many antique shops offer attractive art nouveau or 19th-century 'Historicist' glass items. 'Onion pattern' porcelain is both serviceable and attractive. Apart from the usual souvenirs, embroidered textiles are worth considering, as are hand-crafted puppets. Musical mementoes represent good value: Supraphon's CDs (and even vinyl discs) cover all the important works by Smetana, Dvořák and Leoš Janáček, and there is also plenty of Czech folk music – although that may be an acquired taste.

ANTIQUES

Adamo Galerie
Uhelný trh 6, Staré Město. Tel: 22 55 36.

Antikvariat 'U Karlova Mostu'
Prints, drawings, paintings rare books and maps.
Karlova ulice 2, Staré Město.
Tel: 2422 9205.

Antique Vladimír Andrle
Křižovnická 1, Staré Město.
Tel: 23 11 625.

FOREIGN LANGUAGE BOOKSHOPS

Bohemian Ventures
English language paperbacks.
Náměstí Jana Palacha, Staré Město.
Tel: 2421 3382.

Knihkupectví Melantrich
Maps, books, foreign newspapers.
Na příkopě 3–5, Nové Město. Tel: 2422 7258.

Fashion in the heart of Prague, Na Příkopě

Malostranské knihkupectví
Good selection of books on Prague and art books.
Josefská 2, Malá Strana. Tel: 53 20 01.
Široký dvůr knihkupectví
Delightful little shop with books, postcards, CDs, prints.
Pohořelec 110, Hradčany. Tel: 53 22 65.

DEPARTMENT STORES

Kotva
Five floors of shopping with a huge car park attached.
Náměstí Republiky 8, Nové Město. Tel: 2421 5462. Open until 8pm on Thursday.
Máj
Národní 26, Nové Město. Tel: 2422 7971. Open until 8pm on Thursday.

GLASS AND PORCELAIN

Bohemia Crystal Shop – Jafa
Maiselova 15, Staré Město. Tel: 2481 0009.
Bohemia-Moser
Moser is a quality name in Bohemian glass. Mailing service.
Na příkopě 12, Nové Město. Tel: 2421 1293.
Contrans
Large selection of glass.
Malé náměstí 1, Staré Město. Tel: 26 10 19.
Dana
Slightly cheaper glass and porcelain in an arcade shop.
Národní 43, Nové Město. Tel: 2421 4655.

MUSIC

CD Shop – Studio Matouš
Modest but well-chosen selection of Czech composers on CD.
Palác Kinských, Staroměstské náměstí. Tel: 231 10 39.
Dedika
Czech classical and folk music, CDs.
Celetná 32, Staré Město. Tel: 26 38 31.

Decorated eggs – pretty and inexpensive

Supraphon Store
All types of music.
Jungmannova 20, Nové Město. Tel: 26 33 83.

SOUVENIRS

Souvenir
Cards, folk dolls, Alfons Mucha posters.
Staroměstské náměstí 15. Tel: 231 17 34.

CLOTHES

Adam
Where the presidential suits are made.
Na Příkopě 8, Nové Město. Tel: 261 523.
Liška
Clothing and furs.
Železná 1, Staré Město. Tel: 2422 1928.
Tuzex
Železná 15, Staré Město. Tel: 2422 5967.

PERFUMERY

Tuzex
Perfume and cosmetics.
28 Října 5, Nové Město. Tel: 2421 4737.

> Prices for many items are now approaching those of Western Europe, a trend exacerbated by the introduction of VAT in 1993. You can reclaim VAT on exported items.

BOHEMIAN GLASS

Bohemia means good dumplings, beer and glass. The last two are traditionally inseparable, since the Bohemian glassmakers have always needed constant irrigation as they worked at the glass furnaces. Aristocratic landlords were obliged to brew good beer on their estates to keep their glass-making tenants happy ... which is doubtless the reason why the quality of both glass and beer has never declined.

The tradition of glassmaking in Bohemia goes back at least to 1414, and possibly as far back as the Celts. The process went on deep in the forests, which supplied the wood for the furnaces, and glassmakers moved from place to place as wood stocks were exhausted. In the 17th century the trade separated – makers of raw glass delivered to specialists who cut, engraved and gilded it. At first pedlars took the products all over Europe, but by the beginning of the 18th century 'Bohemian Houses' were marketing the famed Czech glass in 38 European ports, as well as America, North Africa and elsewhere.

Although the glass industry suffered two catastrophic setbacks in its history (at the end of the Napoleonic Wars and when the Sudetenland was lost to Germany in 1938), it has always

Bohemian glassware has been produced in Prague since medieval times. Standards today are as high as ever

bounced back with new techniques. In the Renaissance there were gimmicky innovations (drinking vessels that gurgled or whistled as you drained them); the baroque period was notable for beautiful engraving on a specially strong glass – Bohemian crystal; the 19th century developed brilliantly coloured glazes. The iridised art nouveau glass of the turn of the century is often even more exotic, with its sensuous shapes and metallic lustre.

Almost any antique shop will have a few pieces of decorative historic glass in the window, while the state shops will sell you high quality modern crystal. Prospective purchasers should make up their minds swiftly, however: nothing moves faster than glass in Prague.

Entertainment

ART GALLERIES

There is a lively art scene in Prague and plenty of galleries. Exhibitions are held in many historic buildings such as the House of the Stone Bell and the Kinsky Palace on Staroměstské náměstí.

Galerie Böhm
Contemporary Bohemian glasswork.
Anglická 1, Nové Město. Tel: 2422 2918.

Regular art exhibitions are held at the Emmaus Monastery

Galerie Mladých – U Řečických
Pleasant gallery with a range of affordable art; also catalogues, videos and slides.
Vodičkova 10, Nové Město. Tel: 2421 3648.

Mánes Gallery
The moving spirit behind this gallery was the 19th-century history painter Josef Mánes, who founded an Artists' Association in 1887. The modern building overlooking the Vltava was designed by Otakar Novotný, and incorporates a café.
Masarykovo nábřeží 250, Nové Město. Tel: 29 55 77.

Victoria Art
Popular gallery with a nose for marketable painters (see page 77).
Mostecká 6, Malá Strana.

PHOTOGRAPHY

Prague House of Photography
Exhibitions of contemporary Czech photographers. Photo albums of leading figures such as Josef Sudek (see pages 76–7)
Husova 23, Staré Město. Open: daily 10am–6pm.

THEATRE

Tickets are available at box-offices or through ticket agents: PIS at Na příkopě 20, (tel: 26 40 22), and at Staroměstské náměstí 22, (tel: 2421 2844); or Bohemia at Na příkopě 16, (tel: 2421 5031) or Karlova 8, (tel: 2422 7651).

The language barrier will block most visitors' appreciation of Czech theatre, though some contemporary dramatists are known through translation, notably Bohumil Hrabal and Václav Havel.

Ballet is on offer at the National Theatre (Národní divadlo)

The many small or fringe theatres of Prague put on new work while the National Theatre, Národní 2, tel: 2491 3437) offers classics as well as ballet and opera. Its *Nova Scéna* extension stages the popular multi-media Laterna Magika shows (see pages 86–7), also put on at the Palace of Culture in Vyšehrad.

'BLACK' THEATRE, MIME, PUPPETRY

There is now a lot to choose from and some of the shows are devised with the foreign visitor in mind. The mime tradition of Prague is very sophisticated and the best shows are memorable.

Divadlo na zábradlí
Mime shows as well as straight theatre.
Anenské náměstí 5, Staré Město.
Tel: 2422 1934.

Divadlo Spejbla a Hurvínka
Famous puppet duo created by Josef Škupa.
Římská 45, Vinohrady. Tel: 25 16 66.

Studio Gag
Slapstick comedy by Boris Hybner.
Národní 25, Nové Město. Tel: 2422 9095.

For the permanent collections of the National Gallery see pages 90–1. Contemporary Czech art is currently displayed in the Jízdárna (Prague Castle Riding School) at U Prašného mostu 55, Hradcany (open: Tuesday to Sunday 10am–6pm). However, a Museum of Modern and Contemporary Czech Art is due to open soon in the Veletržní Hall, Holešovice. Call the Jízdárna on 2101, ext 32–32 for the latest information.

CINEMA

The fortnightly English-language newspaper *Prognosis* lists films shown in English (foreign films are otherwise dubbed into Czech – indicated on the poster by a small square). The British Council (Národní 10, Nové Město, tel: 2491 2179) shows films from time to time. Czech cinema is noted for its off-beat humour and deadpan satire. Miloš Forman's early films were examples of this while the most enchanting of the Czech New Wave films was Jiří Menzel's *Closely Observed Trains* (1966).

Music

*D*espite all the political vicissitudes Prague has remained one of the great centres of music-making in Europe (see pages 92–3). Classical music was relatively unaffected by Communism, while folk music was even encouraged, albeit in a sanitised form. On the other hand jazz was the music of protest and subject to periodic police harassment.

CLASSICAL MUSIC

Bohemian musical tradition is rich in baroque composers and 19th-century romantics. A group called Musica Antiqua Praha specialises in baroque works played on original instruments, including items from a rich hoard of unknown music from the archives of a 17th-century Bishop of Olomouc in Moravia, discovered by their director, Paul Kliner. The first privately funded ensemble, the Virtuosi di Praga, concentrates on the music of Mozart. The city also boasts two symphony orchestras: the Prague Symphony Orchestra and the Prague Philharmonic. You can hear live music at the following venues:

BOOKING FOR MUSICAL PERFORMANCES

Tickets for music events can be obtained at the box-offices, or from agencies such as Bohemia at Na příkopě 16, tel: 2421 5031. Bohemia Ticket International, PO Box 534, Praha I handles postal booking for the Prague Spring festival. From mid-April festival tickets are on sale at Hellichova 18. Tickets for the Czech Philharmonic are sold at the Rudolfinum, Náměstí Jana Palacha, no 1 (tel: 2489 3111) and TICKET PRO, Na Příkopě 20, tel: 311 87 70.

St Agnes Convent – setting for music

Opera

Národní divadlo (National Theatre)
Opera and theatre performances.
Národní 2, Nové Město. Tel: 2491 3437.

Státní opera (State Opera)
Opera and ballet.
Wilsonova, Nové Město. Tel: 26 53 53.

Stavovske divadlo (Estates or Tyl Theatre)
Opera, ballet, theatre.

Ovócny trh 6, Staré Město. Tel. 2421 4339.

Operetta and Musicals
Hudební divadlo v Karlíně (Karlin Music Theatre)
Křižíkova 10, Karlín. Tel: 2421 0710.

Concert Halls
Dům umělců (House of Artists/Rudolfinum)
Náměstí Jana Palacha, Staré Město. Tel: 286 03 52/286 02 35.
Smetanova Síň Obecního domu (Smetana Hall of the Community House)
Náměstí Republiky, Nové Město. Tel: 232 25 01.

The **Vila Bertramka** (Mozartova 169, Smíchov, tel: 54 38 93) and **Villa Amerika** (ke Karlovu 20, tel: 29 82 14) hold regular evenings of Mozart and Dvořák arias (see pages 88–9).

Concerts and recitals take place in many of the city's baroque palaces and churches throughout the year.

FOLK MUSIC

Look for *lídova skupina* in the listings. For information tel: 84 81 14 (fax: 684 01 82). The Czech Song and Dance Ensemble performs at the Theatre in Klarov, Nábřeží E Beneše 3 (tel: 2451 1027).

JAZZ

Agherta Jazz Centrum
Cocktails, snacks, CD shop. Cosy.
Krakovská 5, Nové Město. Tel: 22 45 58. Open: daily 9am–midnight.

Luxor Jazz Club
Owned by a group of fans of American jazz. Eclectic range of music.
Václavské náměstí 41, Nové Město. Tel: 2421 4782. Open: daily 8pm–midnight.

PRAGUE FESTIVALS

The Pražské Jaro (Prague Spring) traditionally begins on 12 May with a performance of Smetana's *Má Vlast* (*My Homeland*) and closes on 2 June with Beethoven's *9th Symphony*. In between is a rich programme of opera, choral works, chamber music and recitals given by international stars. In July and August the Prague Cultural Summer features music, dance and theatre. Praga Europa Musica in September presents a programme combining aspects of Czech music with music of another European country.

Folk dancing in the best Czech tradition

Metropolitan Jazz Club
Swing, ragtime, blues in congenial pub-like surroundings.
Jungmannova 14, Nové Město. Tel: 2421 6025. Open: daily 5pm–1am.

Reduta Jazz Club
A great survivor. Dixieland, swing, and modern jazz are all on offer.
Národní 20, Nové Město. Tel: 2491 2246. Open: Monday to Saturday 9pm–midnight.

Nightlife

The ingredients of Prague nightlife are much the same as elsewhere: bars, floor shows and casinos for the middle-aged or older, discotheques and noisy live music venues for the young. A number of places have opened up since the 1989 revolution, and others have been revamped.

A few words of warning are in order. Firstly, the youth scene changes very fast. To keep abreast of the latest fashion you should enlist a Czech friend who is into the scene. Nevertheless, the discos and live venues mentioned below are those that have endured or at least look like enduring. Secondly, a considerable substratum of nightlife (especially around Wenceslas Square) is little more than a shop window for prostitution. The places concerned are typically the seedier looking dance halls with a thuggish bouncer guarding the door. Lastly, the expensive hotel nightclubs and some more modest places insist on 'correct' dress – no jeans, no trainers, tie obligatory.

CABARET, FLOOR SHOWS, VARIETY

Traditional floor shows are mainly the preserve of hotels. Reservations are essential.

Alhambra (Ambassador Hotel)
Variety show, revue.
Václavske náměstí 5, Nové Město. Tel: 2419 3111.

Esplanade
Slick floor show with pretty girls.
Washingtonova 19, Nové Město. Tel: 2421 1715 or 2421 3697.

Forum
Kongressová 4, Vyšehrad. Tel: 6119 1111.

Intercontinental
Náměstí Curieových 5, Staré Město. Tel: 248 8913.

CASINOS

There are casinos in the Hotel Forum (open: 6pm–4am) and the Ambassador (open from 4 pm); also at:

Prague Diplomatic Club
Quality restaurant with French cooking. You can follow your meal with a little flutter on the tables.
Karlova 21, Nové Město. Tel. 26 57 01. Open: noon–midnight.

Casino Hotel Palace
Panská ulice 12. Tel: 2422 5659. Open from 6pm.

DANCING

Bílý Koníček
Located in a Romanesque cellar. Old and new dance hits. A bit seedy, but romantically so.
Staroměstské náměstí 20. Tel: 235 89 27. Open: Tuesday to Saturday 9pm–2am.

Classic Prague Club
Uses the auditorium of a theatre for dancing after performances. Classics of rock and pop. Friendly.

Pařížská 4, Staré Město. Tel: 232 0183. Disco from 10.30pm to 3am.

Lucerna

Dinner, 'Bohemian Fantasy' floor show and dancing. All-in price or separate booking for show only.

Štěpánská 61, Nové Město. Tel: 235 29 09 or 22 30 41. Show starts at 9pm.

Opera Mozart

Dancing on a terrace overlooking the Vltava with a café and theatre.

Novotného lávka 1, Staré Město. Tel: 22 82 34. Open: noon–5am (dancing in the evening).

Peklo

The name means 'hell'. Located in the rock chambers beneath Strahov Monastery. Good restaurant next door.

Strahovské nádvoří 1, Hradčany. Tel: 53 02 15. Open: daily 10pm–4am.

ROCK SCENE AND GIGS

To catch the latest happenings you must study the listings in *Prognosis* (fortnightly), *Golem* (monthly), the *Czech Program* (weekly) or *Přehled* (monthly).

Bunkr

It really was a bunker for Husák and comrades. Czech and foreign bands three weeks in the month, otherwise disco. Café upstairs.

Lodecká 2, Staré Město. Tel: 2481 0475. Open: 6pm–5am.

Rock Café

Hard rock and thrash metal. Sometimes live bands. Bar. T-shirts, discs, posters on sale. Loud.

Národní 20, Nové Město. Tel: 2491 4414. Open: daily 10pm–3am.

Night clubs are mushrooming in Prague

Children

*P*rague is rich in possibilities for keeping children absorbed and contented. There is always plenty going on in Old Town Square (Staromestské náměstí), and the river is another attraction. Caves, museums, puppets and a fun-fair also beckon.

You could begin with the all-action astronomical clock in Old Town Square and follow up with a tour of the sewers (entrance to the right of the clocktower). In summer the square is alive with strolling players, ice-cream vendors and booths selling food and souvenirs. A

Guaranteed to please. Fiacres set out from Old Town Square

miniature train leaves from here and runs along Pařížská, over Cechův most and on to Hradčany. Horse-drawn carriages leave from the square for an hour-long tour of the Staré Město. Another obvious attraction is the river

Cruises start from Palackého most, Rašínovo nábřeží (information on 29 38 03 and 29 83 09) and go both upstream and downstream. The longest round-trip lasts more than seven hours, so check timetables and destinations carefully. On a sunny day fun and exercise may be had by hiring a rowboat from Kampa Island, or the island called Slovanský ostrov or Žofín off Masarykovo nábřeží.

CAVES

The stalactite caves of Koněpruské jeskyně in the karst region around Karlštejn offer an exotic alternative excursion. If the children are too small to attempt the 3 km walk from Srbsko station, an infrequent bus runs between Beroun and the caves. Beroun is 50 minutes by train from Praha Hlavní nádraží (see page 144).

CINEMA

The American Hospitality Centre, Malé náměstí 14 (tel: 2422 9961; open daily 10am–11 pm) is an oasis of American culture (including cable TV, popcorn and Coca-Cola) much frequented by the young. Films for children are shown here on Saturday mornings.

McDONALD'S

There are four in Prague; the central ones are at Vodičkova 15, and on the east side of Wenceslas Square.

MIRROR MAZE

Petřín Hill is ideal for an excursion with younger children, involving the funicular railway from Újezd, the astronomical observatory, the mini-Eiffel Tower and the Bludiště (mirror maze), which never fails to appeal and delight.

MUSEUMS

The two military museums – in the Schwarzenberg Palace on Hradčany Square (up to World War I – rich collection of early weapons) and in Žižkov (post-1914) – will no doubt fascinate older children. The view from the Czech Army Museum in Žižkov is one of the best in Prague. The huge equestrian monument of the Hussite general Jan Žižka, which stands in front of the museum, is in the *Guinness Book of Records* as the largest sculpture in the world.

The Museum of Flying and Cosmonauts is worth the longish trek, but the National Technical Museum is even more guaranteed to be a hit with the young. Some 35,000 objects are displayed, the highlights being the automobiles (including the first to be built in Bohemia), the motorbikes, the astronomical instruments of Tycho Brahe and Kepler and a complete mock-up of a coal-mine.

PUPPET THEATRES

Divadlo minor

Continuous programmes for children.

> The museums mentioned above are covered in detail with full addresses and opening hours on pages 80–3; only aspects of specific interest to children are stressed here. Likewise Koněpruské jeskyně is featured on page 145, Petřín Hill on pages 102–3, Výstaviště on page 146 and cruising on page 144.

Animal magic enthralling young visitors in Prague Zoo

Senovážné náměstí 28. Tel: 2421 3241. Metro to Náměstí Republiky.

Národní divadlo marionet

Matinee performances.

Žatecká 1. Tel: 232 34 29. Trams 17 and 18 and metro to Staroměstská.

VÝSTAVIŠTĚ

Fun-fair, big wheel, planetarium and other attractions.

ZOO

In Troja (7th District). Tel: 6641 0480. Open: daily 9am –7pm, 5pm in winter. Metro to Nádraží Holešovice, then bus 112.

Sport

*T*he most passionately followed sports in Prague are soccer and ice-hockey. The latter famously sparked riots in 1969, when the Czech team humiliated the Soviet Union. This was seen as sweet revenge for the Warsaw Pact invasion of the previous year. Soccer has tended to lose its stars to western clubs and while Czech-born tennis giants such as Martina Navrátilová and Ivan Lendl became world superstars, their careers were made outside their homeland. It remains to be seen whether the Czechs can continue to supply tennis players of world rank.

BILLIARDS

There are eight tables at Strahovský Stadion (Entrance 12). *Tel: 35 72 89, ext 171. Open: 11am–11pm. Bus 176 from Karlovo náměstí.*

Bicycles offer easy access to the countryside

BOWLING

Hotel Forum

Just four lanes (but the 10 types of beer are a compensation).

Kongressová 1. Tel: 6119 1111. Open 2.30pm–12.30am. Metro to Vyšehrad.

FITNESS CENTRES

Aerobic courses, fitness centres and gyms have mushroomed in Prague recently. Two with good reputations are:

Fitcentrum Dlabačov

Sauna, 25m pool and table tennis.

Bělohorská 24. Tel: 311 32 41. Open: Sunday to Friday 1–9pm. Trams 8 and 22 to Bělohorská.

Fitcentrum Natur

Sauna, gym; plus alternative medicine clinic.

Dukelských hrdinů 17. Tel: 37 37 91. Open: Monday to Saturday 10am–9pm. Metro to Vltavská.

GOLF

TS Golf

A nine-hole course where you can rent clubs.

Motol. Tel: 59 66 93. Open: daily 8am–8pm (6pm in winter). Metro to Anděl, then trams 4, 7 and 9 to Motol.

HORSE RACING

Steeplechasing and hurdles take place May to October on Sunday afternoons, trotting all the year round. The main course is at Velká Chuchle (tel: 54 66 10) reached by buses 129, 241, 243 or 245 from Smíchovské nádraží.

ICE HOCKEY
Sparta Praha plays at Výstaviště, Praha 7. Tram 17 to Výstaviště; metro: Nádřaží Holešovice.

ICE SKATING
There are a number of rinks, most with erratic opening times.

HC Praha
Za Lány 6. Tel: 36 27 59. Open: winter only, Thursday 3–5 pm, Saturday 1–3pm, Sunday 3.30–5.45pm. Metro to Dejvická, then tram 26 to Bořislavka.

Zimní Stadión Niklajka
U Nikolajky 28. Tel: 54 09 51. Open: winter only, Saturday and Sunday 1–4pm. Metro to Anděl.

SOCCER
The leading team, as for ice hockey, is called Sparta Praha.

Matches at Letenské sady, Milady Horákové 98. Tel: 37 39 68 (Tuesday and Friday at 6pm from September to June).

SQUASH
There are courts at the Hotel Forum (see Bowling).

SWIMMING
Swimming in the River Vltava is not recommended. There are plenty of pools, but not all are as clean as they should be. There are indoor pools at Plavecký Stadión, Podolská 74, Podolí. Tel: 43 91 52. Opening hours vary. Trams 3 and 17 to Kublov.

Strahov Stadión
Tel: 35 23 84. Opening times vary. Bus 176 from Karlovo náměstí to the last stop.

Tickets and information for sporting events are available from ČSTV, Dukelských hrdinů 13, tel: 37 58 49. Listings from ČEDOK and in *The Month in Prague*. Tickets for soccer and ice-hockey are cheap and available at the gate.

TENNIS
Ring the Czech Tennis Federation (tel: 231 12 70) to book courts.
Štvanice ostrov (open: 6am–dusk). Metro to Florenc.

Sparta in Stromovka has 17 outdoor and two indoor courts.
Tel: 32 48 50. Open: April to October, 6am–9pm. Metro to Hradčanská, then bus 131 to Nadraži Bubeneč.

A future Ivan Lendl? Prague is ideal for tennis enthusiasts

Food and Drink

One of the virility symbols of Communism was the country's per capita meat consumption: under the Husák regime it rose to half a kilo per person per day, nearly the highest in the world. Fat content in food was also encouraged, resulting in one of the lowest life expectancies of 27 European nations.

These unappetising statistics reflect a tradition of Czech cuisine which is strong in the meat department and correspondingly weak in most others. The heavy peasant dishes come from the agrarian heartlands of Bohemia, Moravia and Slovakia. German influence is evident in the various sausages and the ubiquitous pickled cabbage. A welcome relief from red meat is provided by the fresh-water fish, and increased imports of sea fish. Nevertheless, the typical national dish remains roast pork with cabbage and dumplings. Dumplings are the real glory of Bohemian cuisine.

TYPICAL DISHES AND SPECIALITIES

Soups (Polévka) and Entrées (Předkrmy)

Bramborová polévka potato soup
Čočková lentil soup
Fazolová bean soup
Hovězí vývar consommé, bouillon
Kulajda vegetable soup
Chléb bread
Chlebíčky open sandwich
Pražská šunka Prague ham
Pražská šunka s krěnem Prague ham with horseradish
Pražská šunka s okurkou Prague ham with pickle
Tresčí játra cod's liver
Uzený jazyk smoked tongue

Traditional pork dish

Vajíčkový salát egg mayonnaise
Omeleta omelette
Žampióny s vejci mushroom with eggs

Main courses

Biftek s vejcem beef and eggs
Drštky tripe
Guláš goulash
Hovězí beef
Játra liver
Kachna duck
Klobásy sausages
Knedlíky dumplings
Krůta turkey
Kuře chicken
Skopové mutton
Smažený řízek Wiener Schnitzel
Svíčková na smetaně roast loin of beef with cream sauce
Telecí veal
Vepřové se zelím roast pork with sauerkraut

Vegetables (zélenina)

Brambory potatoes
Červená řepa beetroot
Cibule onions
Hranolky French fries

KNEDLÍKY (BOHEMIAN DUMPLINGS)

Dumplings are made from bread, potato dough, soft curd or flour. Necessary accompaniments to meat dishes to blot up the grease and the beer, they also come in more sophisticated guises, the best undoubtedly being fruit dumplings. Few restaurants serve these, but you may be offered them in a Czech home. The most mouth-watering versions are filled with plums, sour cherries or apricots. Here is a recipe for fruit dumplings made from potato dough. Ingredients: 800g of boiled potatoes, 10g of salt, 2 eggs, 100g of semolina, 200g of wholemeal flour, a teaspoonful of cinnamon, fresh fruit for filling.

Peel cold boiled potatoes and grate them. Sprinkle with flour, semolina and salt. Make a well in the mixture for the eggs. Knead into dough and form pancakes. Add the fruit, close the mixture into balls and seal them. Place in boiling water and cook for 20 minutes,. Remove and sprinkle with cinnamon, breadcrumbs fried in butter, or with poppy seeds and sugar. Melted butter or whipped cream also make good toppings.

Kyselé zelí sauerkraut
Lečo ratatouille
Obloha garnish (usually pickled vegetables)
Okurka cucumber
Rajčata tomatoes
Salát salad
Špenát spinach
Zelí cabbage

Fish (ryby)
Kapr vařený s máslem boiled carp with melted butter
Pečená štika roast pike
Platýs flounder
Pstruh na másle trout in melted butter

Dessert (zákusky)
Jablkový závin apple strudel
Omeleta se zavařeninou jam omelette
Palačinky pancakes
Švestkově knedlíky plum dumplings
Zmrzlina ice-cream

Cheese (sýr)
Bryndza goat's cheese in brine
Oštěpek smoked curd cheese
Tvaroh curd cheese
Uzený sýr smoked cheese

Fruit (ovoce)
Banán banana
Borůvky bilberries
Broskev peach
Hroznové víno grapes
Hruška pear
Jablko apple
Kompot stewed fruit
Jahody strawberries
Maliny raspberries
Pomeranč orange
Švestky plums.

Don't count the calories!

Eating Out in Prague

Although the city's restaurants offer some of the most atmospheric medieval and baroque interiors in Europe, it cannot be said that the food and service have always lived up to the surroundings. Matters are now noticeably improving. Overcharging, which had become endemic, is energetically combated by the Czech Commercial Inspectorate. In 1992 they found that over 70 per cent of restaurants inspected in Prague's first district were inflating their bills. The moral for visitors is: check the waiter's 'arithmetic' every time! If it proves to be in order and if the service has been halfway acceptable, a tip of 10 to 15 per cent is usual in the better establishments, a few crowns elsewhere.

Reservations are vital for restaurants, especially in the high season. The most popular places may need to be booked as much as a week in advance, although two to three days is usually adequate. If you have not booked anywhere you may have to fall back on a beer cellar, a café or a fast food outlet (see pages 174–5).

TYPES OF EATING HOUSE

The choice of establishments falls into three main categories: *restaurace* (restaurant), *vinárna* (wine-bar or wine cellar) and *pivnice* (beer cellar). The last named usually offer pub food, although that should be regarded in most cases simply as an accompaniment to the serious business of beer consumption.

Creeping gentrification has meant that the distinctions between different types of hostelry have become blurred: in particular a number of *vinárnas* are now luxury restaurants in all but name.

Because of inflation and changes of ownership the following guide to prices can only be an indication. Average meal prices per head are implied, exclusive of drink.

K up to 150kč
KK up to 275kč
KKK up to 450kč
KKKK over 450kč.

Prices can be expected to continue rising, especially in the sort of place largely frequented by foreigners (already beyond the pockets of all but a tiny minority of Czechs).

The price of a glass of wine varies, but 40kč for a quarter litre is not unusual. Beer prices show even greater disparities – anything from around 15kč in a traditional pub to 30kč or more in a popular tourist spot such as U Flecků.

One of Prague's best known restaurants: 'At the Spider'

The national dishes are exceptionally rich and filling

BOHEMIAN CUISINE

Jihočeské Pohostinství (South Bohemian Inn) KK

Typical South Bohemian fare with much on offer to delight dumpling fanciers.

Na příkopě 17. Tel: 2421 0661. Open: daily 11am–11pm. Metro to Můstek.

U čižků KKK

Good quality and good value. Duck, goose and a rare wonder on a Prague menu: fruit dumplings.

Karlovo náměstí 34. Tel: 29 88 91. Open: Monday to Sunday noon–3.30pm, 5–10pm. Metro to Karlovo náměstí.

U pavouka (The Spider) KKKK

The restaurant is in one of the oldest houses in Prague. Good veal and duck, friendly service.

Celetná 17. Tel: 2481 1436. Open: daily noon–3pm, 6pm–midnight. Metro to Náměstí Republiky.

U tří pštrosů (The Three Ostriches) KKKK

In this former Renaissance house (also an exclusive hotel) you can sample traditional Bohemian dishes washed down with Pilsner beer and Moravian wine.

Dražičkého náměstí 12. Tel: 2451 0761. Open: daily noon–3pm, 6–11pm. Trams 12 and 22 to Malostranské náměstí.

U zlaté hrušky (The Golden Pear) KKKK

Attractive carved peasant furniture is a feature here. Duck, venison and goose liver are recommended.

Nový Svět 3. Tel: 53 11 33. Open: daily 11.30am–3 pm, 6.30pm–midnight. Tram 22 to Brusnice.

INTERNATIONAL CUISINE

David KKK
Small and chic with foodstuffs imported from Germany. Good salad and poultry.
Tržistě 21. Tel: 53 93 25. Open: daily noon–3pm, 6–11pm. Trams 12 and 22 to Malostranské náměstí.

Nebozízek KKK
Reached by the funicular railway on Petrín, this is celebrated as much for the wonderful view as for the rather average cuisine.
Petřínské sady 411. Tel: 53 79 05. Open: daily 11am–6pm, 7–11pm. Trams 6, 9, 12 and 22 to Újezd, then via railway.

Art nouveau and French cuisine in the Obecní Dům

Opera-Grill KKKK
A small and intimate establishment offering delicacies hard to come by in Prague, such as snails and lobster.
Karolíny Světlé 35. Tel: 26 55 08. Open: daily 7pm–2am. Metro to Národní třída. Trams 6, 9, 18 and 22 to Národní divadlo.

U zátiší (The Quiet Corner) KKK
Nouvelle cuisine comes to Prague! So few restaurants in the city serve decent fresh vegetables, it is worth it for these alone.
Liliová 1. Tel: 2422 8977. Open: daily noon–2.45pm, 6–10.45pm. Metro to Národní třída.

FISH RESTAURANTS

Český Rybářský Svaz KK
The home of the Fishermen's Club, where the wives do the cooking. The fish comes from ponds in Central Bohemia.
U sovových mlýnů 1. Tel: 2451 0972. Open: daily 11am–11pm. Trams 6, 9, 12 and 22 to Újezd.

Reykjavík KKK
An excellent addition to the Prague scene, selling fast-frozen Icelandic fish that tastes absolutely fresh: salmon, shrimp, cod, haddock. Fish soup a speciality. Service excellent; no reservations.
Karlova 20. Tel: 2422 9251. Open: daily 11am–11pm. Metro to Staroměstská.

Vltava K
Freshwater fish and fruit dumplings. Said to have the best pancakes in town.
Rašínovo nábřeží. Tel: 29 49 64. Open: daily 11am–10pm. Trams 3, 7 and 17 to Rašínovo nábř.

U vojáčků K
Trout, carp, catfish, pike and perch. Pilsner from the barrel. Menu in Czech.
Vodní 11. Tel: 53 56 68. Open: Monday to Saturday noon–midnight. Metro to Anděl, then trams 6, 9 and 12 to Kinského Zahrada.

GAME RESTAURANTS

Myslivna KKK

Traditional game restaurant with pleasant atmosphere.

Jagellonská 21. Tel: 627 02 09. Open: daily 11am–4pm, 5pm–midnight. Metro to Flora.

U lípy (At the Lime-Tree) KKK

A wide range of game including hare, quail and wild boar is offered. The wine list has more than one hundred items.

Plzeňská 142. Tel: 52 29 27. Open: daily 11am–3pm, 6pm–midnight. Trams 4 and 9 to Krematorium Motol.

Zdar

Good food, good service, good value.

Vinohradská 164. Tel: 6731 0898. Open: Sunday to Friday 11am–10pm. Metro to Flora.

ETHNIC RESTAURANTS

CHINESE

Dlouha Zěd ČChang-ČCheng (The Great Wall) KK

Prague's only absolutely authentic Chinese restaurant – or so they say.

Marie Pujmanové 10. Tel: 692 23 74. Open: Tuesday to Sunday 11am–3pm, 4.30–11pm. Metro to Pankrác.

Fenix KK

This has all the old favourites of the Chinese kitchen, including delicious steamed dumplings.

Vinohradská 88. Tel: 25 03 64. Open: daily 11am–3pm, 5.30–11pm. Metro to Jiřího z Poděbrad.

Zlatý drak KK

Eighty-seven dishes, and chefs imported from China to cook them.

Anglická 6. Tel: 2421 8154. Open: daily 11.30am–3pm, 6.30–10.30pm; Sunday 6.30–10.30pm only. Metro to Náměstí míru.

FRENCH

U malířů KKKK

Expensive but gastronomically superb.

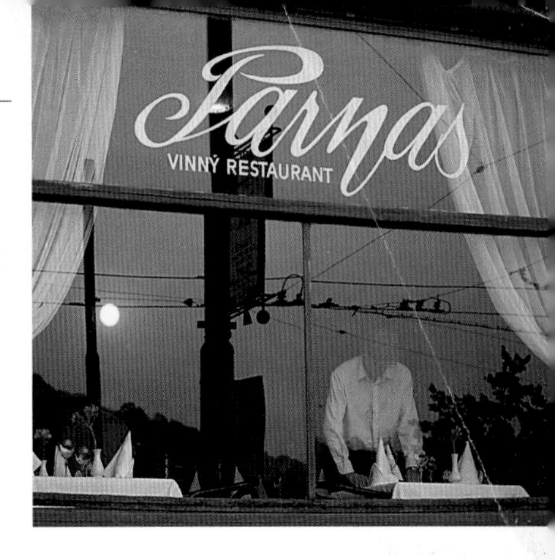

Maltézské náměstí 11. Tel: 2451 0269. Open: daily 7pm–midnight. Trams 12 and 22 to Hellichova.

INDIAN

Mayur KKK

The original Indian managers and staff have long since gone and the food is not all that authentic any more; but the bread and curries are acceptable.

Štěpánská 61. Tel: 2422 6737. Open: daily noon–11pm. Metro to Můstek.

INDONESIAN

Sate K

Good and cheap snack restaurant with typical Indonesian dishes.

Pohořelec 3. Tel: 53 21 13. Open: 10.30am–8pm. Tram 22 to Památník písemnictví.

ITALIAN

Principe Caffe-Ristorante KKKK

Good genuine Italian food and wines.

Anglická 23. Tel: 25 96 14. Open: daily noon–10.30pm. Metro to náměstí Míru.

RUSSIAN

Moskva KKK

Bortsch, caviar, together with Armenian, Georgian and other regional specialities.

Na příkopě 29. Tel: 2422 9268. Open: daily 11am–1am. Metro to Můstek.

Drink

The two major wine-growing areas of former Czechoslovakia are around Břeclav in Southern Moravia and Pezinok in Western Slovakia. Both regions are near enough to Prague for pleasant weekend excursions; these can be a lot of fun (and pretty alcoholic) during the wine harvest season from mid-September to late October. The picturesque Moravian *sklípeks* (wine cellars) that border the vineyards offer opportunities for wine-tasting.

Neither the Czech lands nor Slovakia can truthfully be described as producing classic wines, but the best of them are pleasantly drinkable. The consensus is that Moravian wine outshines the rest and many of the Prague cellars specialise in their products. Good wine is also made at Mělník which is near enough to the capital for a lunch or dinner outing (see page 139).

Wines to look out for

In common with neighbouring Hungary and Austria, the white wine is generally more favoured than the red and almost invariably drunk within a couple of years of the harvest.

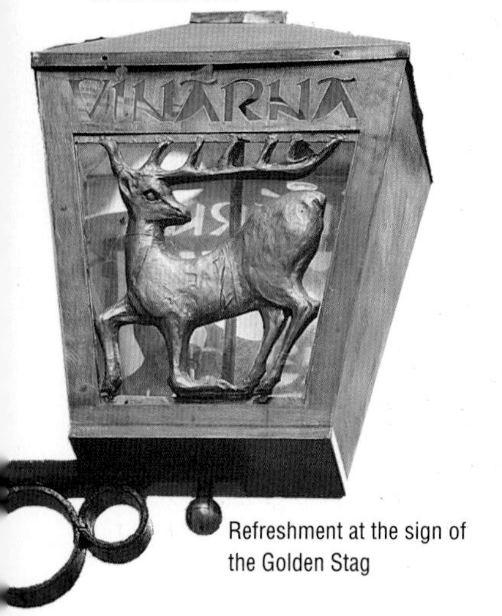

Refreshment at the sign of the Golden Stag

Sauvignon, which is aromatic with a flavour of ripe peaches, is often an exception to this general rule – the best is said to come from Velké Pavlovice in Moravia. *Ryzlink rýnský*, claimed to be the 'king of wines and the wine of kings', has a bouquet of lime blossom and is good with fish. *Rulandské bílé* is more full-bodied and thus often likened to a Burgundy. Connoisseurs of Austrian wine will warm to the *Veltlínské zelené*, the fresh, somewhat acidic 'Grüner Veltliner'. Czechs do not mind drinking white wine with roasts, but for this they will probably choose a *Neuburské* (Neuburger) with its slightly smoky taste.

Müller Thurgau is good with fish or veal, while *Silván* is suited to paté and chicken. A fine Moravian riesling with limited production is *Bzenecká lipka*, often drunk with grilled meat. Of the reds *Rulandské červené* has some of the characteristics of a Burgundy, while the velvety *Vavřinecké* wins praise from red wine enthusiasts.

VINÁRNY

Prague has a vast number of 'wine cellars' or 'wine restaurants' to choose from. Here are some recommendations.

Lobkovická vinarná KKKK

A place to sample Mělník and South Moravian wines. Exclusive and not cheap. Reservation essential.

Vlašská 16–17. Tel: 53 01 85. Open: daily noon–3pm, 6.30pm–midnight. Trams 12 and 22 to Malostranské náměstí.

U mecenáše (Maecenas) KKKK

Up-market establishment in a 400-year-old cellar. Famous for its steaks. Reservation essential.

Malostranské náměstí 10. Tel: 53 38 81/4. Open: daily 5–11.30pm. Weekends, evenings only. Trams 12 and 22 to Malostranské náměstí.

Paris Praha (Au Rendez-vous de Paris) KK

For those gasping for a drop of French wine.

Jindřišská 7. Tel: 236 75 45. Open: Monday to Friday 8.30am–8pm, Saturday 8.30am–2pm. Metro to Můstek.

U tří grácií (The Three Graces) KK

Sixteen varieties of Valtické wine from southern Moravia are on offer, of which the Grüner Veltliner is much praised. Views of the river and Hradčany from the terrace.

Novotného lávka 5. Tel: 2422 9106. Open: daily noon–midnight. Metro to Staroměstská.

Rotisserie KK

A 'wine restaurant' with first class international cuisine. Reservation essential.

Mikulandská 6. Tel: 2491 2334. Open: daily 11.30am–3.30pm, 5.30–11.30pm. Metro to Národní třída.

U zlaté konvicé (The Golden Jug) KKK

Excellent Italian and Rhine riesling (of the whites) and Frankovka (of the reds).

Melantrichova 20. Tel: 2422 7885. Open: daily 6.30pm–12.30am. Metro to Můstek.

Vinárna Paukert KK

Wine from Mělnik. Cold buffet.

Narodní 17. Tel: 2423 0031. Open: Monday to Friday 9am–7pm. Trams 6, 9, 18 and 22 to Národní divadlo and metro to Národní třída.

PIVNICE (Beer Halls)

U Betlémské kaple K

Bohumil Hrabal wrote lyrically about the *Velkopopovický kozel* beer served here 'with its marvellous head ... like whipped cream'. The pub is good value and offers tasty food.

Betlémské náměstí 2. Tel: 2421 1879. Open: daily 9.30am–9pm. Metro to Národní třída.

U dvou Slunců (The Two Suns) KK

This baroque house where the writer Jan Neruda once lived is now a congenial small restaurant serving three sorts of beer from the barrel. Traditional Czech cuisine.

Waiters are much in demand at U Fleků

> **CAFÉ CULTURE**
> It is a sad fact that café culture has decayed in Prague. Too often one is served an indefinable brown substance or an unappetising 'Turkish' coffee. An increasing number of places do offer espresso coffee or *vídeňská káva* (Viennese coffee with whipped cream). Most of the coffee-houses listed are worth visiting for the nostalgic atmosphere at least. All K rating.

Nerudova 47. Tel: 53 89 24. Open: 11am–midnight. Trams 1 and, 22 to Malostranské náměstí.

U Fleků KK

The sweetish black ale, made since 1843 to a Bavarian recipe, is brewed and sold only on the premises. It is best to sit in the garden – brimming mugs arrive automatically and there are normally one or two hot dishes of the day to choose from.

Křemencova 11. Tel: 2491 5119. Open: daily 9am–11pm. Trams 3, 9,14 and 24 to Lazarská. See page 123.

U Kalicha (The Chalice) K

This was *The Good Soldier Švejk*'s favourite watering hole; the place is very touristified but the food is acceptable and the Pilsner more so. Reservations advisable.

Na bojišti 12. Tel: 29 07 01. Open: 11am–3pm, 5–11 pm. Metro to I P Pavlova.

KAVÁRNA (Cafés)

Arco

Old-style literary café with distinguished tradition.

Hybernská 16. Open: daily 7am–11pm. Metro to Náměstí Republiky.

Café Evropa

Indifferent menu but superb art nouveau interior and nostalgic musical duo in the evenings.

Václavske náměstí 29. Open: daily 7am–11pm. Metro to Muzeum or Můstek.

Café Savoy

Enjoy the painted neo-Renaissance ceiling and Viennese interior of this beautifully restored café.

Vitezná 1. Open: daily 9am–11pm. Trams 9 and 22 to Újezd.

Malostranská kavárna

An attractive and cosy 19th-century café equally beloved of residents and visitors.

Malostranské náměstí 28. Open: daily 7am–11pm. Trams 12 and 22 to Malostranské náměstí. Currently closed for restoration.

Obecní dům (The Community House)

Another fine art nouveau interior. A pleasant place to write your postcards home.

Náměstí Republiky. Open: 7am–11pm. Metro to Náměstí Republiky. Currently closed for restoration.

Slavia

The most famous of the writers' (and dissidents') cafés; its future is uncertain.

Národní 1. Tel: 2422 9248. Open: daily 8am–11pm. Trams 6, 9, 18 and 22 to Národní divadlo. Currently closed for restoration.

FAST FOOD/SNACKS/SIMPLE MEALS

Baltic Grill K

Good value if you choose the cheaper fish on the menu.

Václavská náměstí 43. Open: Sunday to Friday 11am–11pm. Metro to Muzeum or Můstek.

In the dozens of ale-houses in Prague you can encounter a wide range of beers, food and prices. Most pubs are tied to the products of one brewery – Pilsner Urquell, Budvar or more often the beers from one of the breweries in Prague itself.

U Bindrů K

Buffet opposite the Old Town Hall.

Staroměstské náměstí 26. Open: daily 10am–8pm. Metro to Staroměstská.

McDonald's

The nicest of Prague's four (the one on Václavske náměstí is always crowded).

Vodickovà 15. Open: daily 8am–11pm. Trams 3, 9, 14 and 24. Metro to Můstek.

VEGETARIAN

Country Life

Health food store with take-away service.

Melantrichova 15. Open: Monday to Thursday 9.30am–6.30pm (Sunday noon–6pm), Friday 8.30am–3pm. Metro to Můstek.

Potato pancakes are irresistible

BEER CULTURE

Beer, to the true-born Czech, is not so much a drink, more a way of life. It is nearly impossible to drink an unpalatable brew in Prague and aficionados of brands are as passionately divided in their enthusiasm as the supporters of the city's two football clubs, Sparta and Slavia.

The distinctive bottom-fermented *Plzeňský prazdroj* (Pilsner Urquell) was first produced in the Bohemian town of Plzeň in 1842. Its great rival is *Budvar* from České Budějovice (Budweis). Flavoured with the hand-picked 'red' hops of Bohemia, the secret of these tipples (as of Scotch whisky) is the soft water used in the brewing. A high carbon dioxide content ensures a fine, flowery head: to test the quality, Praguers put a matchstick into the foam. The contents are satisfactory if it stays erect for at least ten seconds. Czech breweries have mercifully stuck to traditional ingredients and methods, which means that the beer-drinker is spared the characterless chemical fizz cynically foisted on western consumers.

The seriousness of these drinking matters may be seen from the founding of a Party for the Friends of Beer after the Velvet Revolution. Its members appointed themselves guardians of beer quality in the capital, which naturally involved them in the onerous duties of rigorous testing and consumption. Unfortunately the party has been unable to prevent a scandalous rise in prices, although connoisseurs will whisper to each other the names of a few places that still serve pils at 12kč or less. The discovery that one well-known hostelry had recently raised its prices to an unheard-of 50kč per glass provoked national outrage. A country whose president once worked as a brewery hand does not take kindly to profiteering with a commodity so close to a Czech's heart.

Pilsner Urquell, one of the most distinctive Czech beers

In Prague, beer is a way of life

Budvar is not to be confused with its American counterpart, Budweiser

Hotels and Accommodation

With only 10,000 hotel beds serving ever more visitors the Prague hotel industry was for long an unscrupulous businessman's dream: as a Socialist monopoly it could disregard standards of service and exploit captive guests with high prices. It would be nice to report that the Velvet Revolution of 1989 had changed all that, but such structural changes require time and the full benefits of competition will take a while to emerge.

Many of the current problems are transitional: ownership of some hotels has been cast into doubt by the restitution law; others have been closed for lengthy renovation. When the dust has settled, there will undoubtedly be a better and more appealing choice of rooms, although prices will naturally be higher still.

PRICES

Most hotels are now switching to the western star system of grading, although it will be some time before service and accommodation fully meet western standards for each category. In general, hotels are still expensive for what they offer. Private accommodation now provides a pleasant and economic alternative.

The following is an indication of what one might expect to pay for a double hotel room in Prague at the time of writing. Inflation and renovation costs will certainly increase prices. Breakfast is not always included and hotels in the bottom two categories may have rooms without en suite bathrooms. Below 5 stars the star rating requires flexible interpretation. Some hotels may still demand payment in hard currency.

5 star 7,000kč or above
4 star 4,000– 7,000kč
3 star 2,500–4,000kč
2 star 1,500–2,500kč
1 star up to 1,500kč

Private accommodation is usually better value than a modest hotel, although in some cases you may be sharing facilities with a family and in others the hospitality may overwhelm the faint-hearted. Prices start at around 350kč per person. Rooms

The Diplomat Hotel has businessmen in mind but all are welcome

in the seasonally rented college dormitories are cheaper still at around 150–200kč per person.

LOCATION

The hotels around the core of old Prague (Staré Město, Nové Město, Malá Strana) are almost all expensive. A few new luxury hotels have been built further out, but all of them have good communications to the centre. Prague retains a fair mix of residential and business properties around and in the old heart of the city, which means that you can usually find private accommodation in the centre (be sure to check on the map exactly where the room on offer is located).

Motels, camping sites and youth hostels are on the periphery or just outside the city. However, main arteries with buses or trams run close to most of them.

BOOKING

Trying to book a hotel direct can be a frustrating business and is not always reliable. If you book from abroad the state travel agency Čedok still monopolises the business.
Čedok, London: 17–18 Old Bond Street, London W1X 4RB. Tel: 071–378 6009.
Čedok, New York: 10 E 40th Street, NY 10016. Tel: 212–689 9720.
Travellers may still sometimes be told that Prague is 'fully booked' – apparently for a whole season. This could be technically true, in the sense that Čedok's hotels, which are in the middle to upper price range, are often block-booked a year ahead by package tour operators. Of course there will still be accommodation available elsewhere.

Reaching for the skies, the Forum Hotel in Vyšehrad

Thomas Cook

Travellers who purchase their travel tickets from a Thomas Cook network location are entitled to use the services of any other Thomas Cook network location, free of charge, to make hotel reservations.

Restaurant at the Evropa Hotel

BOOKING AGENCIES IN PRAGUE

If you have arrived in Prague without a booking (or wish to change the hotel you have), there are now a lot of agencies that can help you find accommodation. The following list is by no means exhaustive.

Ave

Conveniently situated at the airport and the main railway station. Queues in high season.

Hlavní nádřaží (Central Station). Tel: 2422 3226/3521. Open: daily 6am–11 pm.

Čedok

You can check via their computer whether any hotels have a room currently vacant. Outside the office people offering private accommodation congregate in the season.

Panská 5. Tel: 2421 3495. Open: Monday to Friday 9am–9.45pm, Saturday 8.30am–6pm, Sunday 8.30am–4.30pm. Also at Na přikopě 18. Tel: 2419 7111.

CKM-SSM

Youth oriented, offering cheap accommodation. Discounts and priority booking for youth hostellers.

Žitná 2. Tel: 2491 0251. Open: daily 9am–6pm.

Hello Ltd

Some nice private rooms and apartments for rent.

Senovážné náměstí 3. Tel: 2421 2741. Open: daily 9am–10pm.

IFB

Private rooms between March and October.

Václavské náměstí 25. Tel: 2422 7253. Open: Monday to Friday 9am–1pm, 2–6pm, Saturday 9am–4pm, Sunday 9am–2pm.

Pragotour

Cheaper hotels and private rooms, not usually situated very centrally.

U obecního domu 2. Tel: 2481 1330. Opening hours vary according to season, but generally 8am–8 pm from Monday to Friday 9am–6pm Saturday (to 3pm on Sunday).

Toptour
Upper end of the market for rooms and flats.
Rybná 3. Tel: 232 10 77. Open: Monday to Friday 9am–8pm, Saturday and Sunday 10am–7pm.
Uniset
Spálená 51 (tel: 2491 0113) and Havelská 15 (tel: 26 36 43). Open: Monday to Friday 9am–5.30pm.

HOTELS

Luxury Hotels

The shortage of hotel beds at the luxury end of the market was already being remedied under the Communist regime, with an eye to the foreign business traveller. The most recently built of the luxury hotels are the **Atrium** (tel: 284 11 11) and the **Forum** (tel: 419 01 11), while the **Intercontinental** (tel: 280 01 11) is long established. For fitness fanatics there is the **Club Hotel Průhonice** (tel: 643 65 01), just outside the city, with tennis courts, squash courts, bowling, a gym and riding facilities.

Traditional Hotels

Sadly, only a few of the old style hotels have retained real character and authentic décor. The best preserved of the art nouveau hotels are the **Evropa** (tel: 2422 8118) and the **Pařiž** (tel: 2422 2151) but they are difficult to book. A truly atmospheric little hotel is **U tří pštrosů** (tel: 2451 0779) in a Renaissance house at the Malá Strana end of Charles Bridge. It also boasts a well-known restaurant (see page 165). An interesting new venture at the top end of the market is the **Ungelt** (tel: 2481 1330) in a Gothic house behind Staroměstské náměstí: six suites are in operation, with more planned.

There are a number of other traditional hotels that tend to be clustered round the centre: the well-liked **Ambassador** (tel: 2419 3111), the 1930s-style **Juliš** (formerly the Tatran, tel: 2421 7092) and the **Zlatá husa** are all on Václavské náměstí.

Botels
Staying on the Vltava sounds appealing, but the reality may be less enchanting than the idea. There are three 'botels': the **Admirál**, the **Albatros** and the **Racek**.

Pensions/small hotels

Pensions and cheaper modern hotels are mostly some way from the centre; many are in the 6th and 8th districts. Entrepreneurs are beginning to open new ones.

Seasonal hotels

Some college dormitories are turned into seasonal hotels which can be booked through **Uniset**.

Youth Hostel

To book the youth hostel go to the CKM, Žitná 10 (with your membership card). Tel: 29 29 84.

On Business

The new Czech Republic is looking to the west for business connections, in particular to members of the European Community. Many of the centralised structures that monopolised economic activity under Communism (when 80 per cent of trade was with Eastern Bloc countries) are still around; however they are either in the process of being privatised or are attempting to adapt themselves to life in a competitive market environment.

Bohemia and Moravia offer the most attractive investment possibilities for the future – in tourism, brewing, services and other areas. The Czech crown was reasonably stable in the early 1990s, assisted by tight monetary policies pursued by the Klaus government. However, the impact of collapsing markets in the east, together with root and branch structural reforms, precipitated an economic recession. On the positive side, a small surplus on the balance of trade was achieved in 1991 and the Czech economy is underpinned by IMF support.

BUSINESS HOURS

Offices and ministries in Prague are open 8am–4pm, Monday to Friday. Some offices have staggered hours, starting at 7am, 7.30am or 8am and ending at 3pm or 4.30pm. Banks are generally open 8am–2pm, Monday to Friday, but some in the city centre have longer hours and close for an hour at lunchtime. The first shift in factories is 6am–2pm, the second shift 2–10pm and the third (if worked) 10pm–6 am. For shopping hours see **Practical Guide**, page 187.

CONFERENCE CENTRES AND TRADE FAIRS

The **Diplomat Hotel** (Evropská 15, tel: 2439 4111, fax 34 17 31) has conference and other facilities for businessmen. Information is also available from **Conference Czechoslovakia Ltd** (tel: 311 83 26, fax 311 80 44). The branch of the state tourist agency **Čedok** dealing with congresses and symposia is at Panská 5 (tel: 2423 0135).

The main venue for trade fairs is the **Výstaviště** complex in Holešovice (Metro line C to Nádraží Holešovice, trams 5, 12 and 17 to Výstaviště). Smaller shows (eg computers and software) are sometimes held at **U Hybernů** (Metro to Náměstí Republiky).

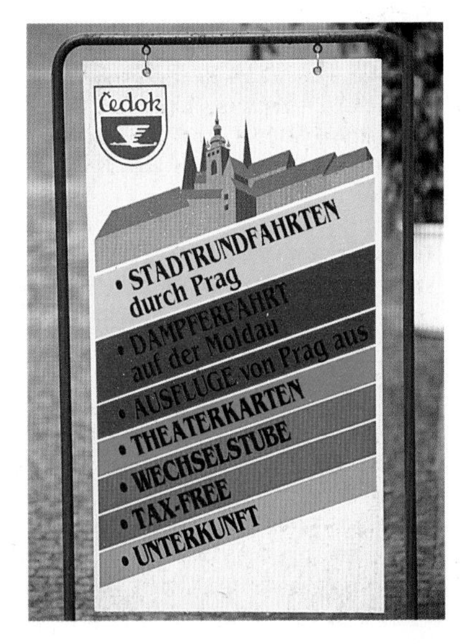

ETIQUETTE

Wearing a suit is not yet obligatory for the businessman here, so a more casual mode of dress should not be taken as implying lack of seriousness. Czechs are punctilious in matters of formal courtesy, always shaking hands on meeting and at leave-taking. Punctuality in the keeping of appointments is also important.

MONEY

The *Prague Post* (a weekly published on Tuesdays) gives the current exchange rates and has commerce-oriented articles in its finance section. The Czech *koruna* (crown) has been made internally convertible, so that companies can obtain the hard currency they need on application to the National Bank. The aim is eventually to make it externally convertible also.

SECRETARIAL AND INTERPRETER SERVICES

Typing

In the Černá ruzé passage at Panská 4 (1st floor – tel: 22 14 95) typing facilities are available – rates vary according to whether you do your own typing, need immediate delivery or have 12- or 48-hour service.

Translation and interpreting

The Prague Information Service (PIS, Za Poříčskou bránou, tel: 26 58 23/4) can help with finding translators. For interpreters, tel: 26 40 94 or 26 68 00. Other organisations offering interpreting services include:

Art-Lingua, Myslikova 6, tel: 29 37 41, fax: 29 55 83.

Babel Service, Palác kultury (close to Forum Hotel), tel: 692 67 41.

TAP, Na příkopě 10, tel: 2421 1443, 2422 6629.

Temporary exhibitions are held at the U Hybernů Centre

Telefax/telegram/telephone

Available at the Main Post Office at Jindřišská 14 (open 24 hours). See also **Telephones**, page 189.

Courier

An international courier service (DHL) is available at the Forum Hotel, Kongressová 1 (tel: 6119 1238/9). For a local messenger tel: 311 63 98, weekdays 8am–5 pm.

Photocopying

There are machines at the Máj Department Store (Národní 26), at Vladislavova 9 and Štěpánská 36. Also in some of the arcades off Václavské náměstí.

Office supplies

Jindřišská 7 (tel: 2421 6924); Kotva Department Store, Náměstí Republiky 8, tel: 235 00 01 or 286 11 11.

Practical Guide

CONTENTS

ARRIVING

Formalities

A valid passport and, in the case of travellers from Australia, New Zealand, South Africa and Canada, a visa (valid for three months) are required. Travellers who require a visa should obtain them in their country of residence, as it may prove difficult to obtain them elsewhere.

By air

Ruzyně airport is 16km west of the city. Czechoslovakian Airlines (ČSA) runs a half-hourly bus service from the airport to their terminal at Revoluční 25 from 5.30am to 7pm. A Čedok bus provides a shuttle service to leading hotels four times a day. Czech Airhandling's mini-bus will drop you anywhere in central Prague for about 200kč and they operate until late evening. The airport has facilities for post, telephone, currency exchange (24 hours), a car rental desk and an accommodation bureau. Flight enquiry number is (2) 36 78 14 or 36 77 60. Prague is served by the leading European and American carriers including Air France, Austrian Airlines, Alitalia, BA, Delta, Lufthansa, SAS, Swissair and United.

Any Thomas Cook Network location will offer airline ticket re-routing and revalidation free of charge to MasterCard cardholders and to travellers who have purchased their travel tickets from Thomas Cook.

By bus

A direct service from London to Prague is operated by Kingscourt Express. It leaves Victoria Coach Station at 7pm on Wednesdays and Saturdays, arriving 24 hours later at Praha-Florenc (Metro station: Florenc).

By car

The territory of the Czech Republic has border crossings with Germany, Poland, the Slovak Republic and Austria. All except a few on minor roads are open 24 hours.

By rail

From London to Prague takes 24 to 30 hours according to route (the most direct route is from London Victoria via Dover, Ostend, Brussels and Frankfurt). Information from British Rail European Travel Centre, Victoria Station, London SW1V 1JY (tel: 071–834 2345). International services run direct to Prague from Vienna (Franz Josefs station), and Berlin, Frankfurt, Munich and other German cities. Trains arrive in Prague at Wilsonovo nádraží (Central Railway Station) or further out at the Nádraží Praha Holešovice. Both are on the Metro (line C). InterRail cards are valid for travel in the Czech Republic.

CAMPING

For camping information contact **UAMK**, Mánesova 20, Praha. Tel: 74 74 00 or 2422 1635 (Monday to Friday).
Sportourist, Národní 33, Praha 1.
There are camping sites at the following locations:
Aritma Džbán, Nad lávkou 3, Vokovice. Tel: 38 90 06. Open: all year round.
Caravancamp, Plzeňská, Motol. Tel: 52 16 32. Open: April to October.
Karavan Park, Císařská louka 599. Tel: 54 09 25. Open: April to September.
Kotva Braník, U ledáren 55. Tel: 46 17 12, 46 13 97. Open: April to October.
Na Vlachove, Rudé armády. Tel: 6641 0214. Open: April to October.
Triocamp, Ústecká. Tel: 6641 1180. Open: all year round.

CHILDREN

Children under 12 travel free on public

Keeping up with the times

transport in Prague. They must be under five to travel free on Czech trains, but five to ten-year-olds pay half fare. The availability of disposable nappies and convenience baby food is improving. There are childcare clinics at the hospital complex at Na Homolce, Roentgenova 2, Smíchov (see **Health**), tel: 5292 2043 and Ke Karlova 2, tel: 2491 1444.

CLIMATE

See page 17.

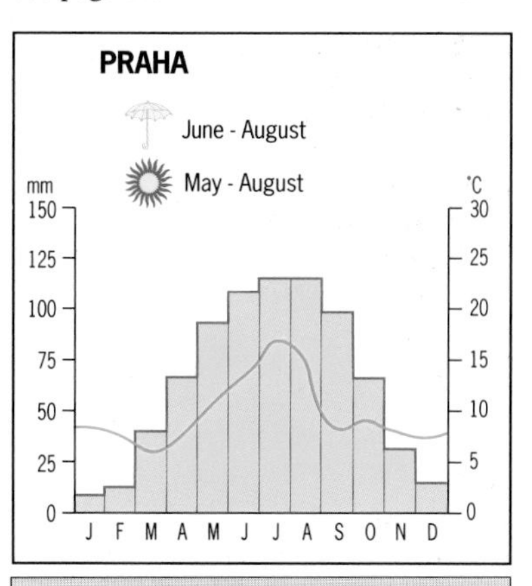

WEATHER CONVERSION CHART
25.4mm = 1 inch
°F = 1.8 × °C + 32

Pretty as a picture

CONVERSION TABLES

See tables opposite.

CRIME

Petty crime (especially pickpocketing and breaking into cars) has greatly increased since the collapse of the Communist regime. Trams on tourist routes (eg no 22) are worked by pickpocket gangs. No valuables should be left in hotel rooms or in cars.

Crime should be reported at the main police station at Bartolomějska 6, Staré Město, Praha 1 (tel: 2413 1111). The emergency telephone number for the police is 158.

Keep a photocopy handy of your passport details, credit card numbers and the like in case of theft.

CUSTOMS REGULATIONS

It is permissible to bring into the Czech Republic free of duty all personal effects plus 250 cigarettes, two litres of wine, one litre of spirits and half a litre of perfume.

Gifts up to the value of 1,000kč are also duty free. The usual prohibitions apply to drugs, firearms etc, and Czech currency over 100kčmay not be taken in or out. The Customs Office is at Havlíčková 11, tel: 2421 4259.

DISABLED TRAVELLERS

Prague is slowly becoming user-friendly for the disabled. It is worth contacting Metatur at Stefánikova 59, Praha 5, (tel: 54 95 63, 55 10 52, a tour agency that specialises in meeting the needs of the disabled.

DRIVING

Driving in Prague

Driving in the centre of Prague is not easy. One-way streets and lack of parking create difficulties and it is easy to stray into zones restricted to public transport and permit holders (eg Wenceslas Square). On the spot fines are levied from offenders. If you are towed away, ring 158 to find out where your motor is held (there are three car pounds, all far from the centre).

Parking

Most parking slots around the centre are for residents only. Whenever possible park on the outskirts of the city and continue your journey by public transport.

Breakdown

The 'Yellow Angels' provide a breakdown service and can be contacted on 02/123. It is advisable to have a comprehensive insurance to cover all eventualities and repatriation. For the 24-hour repair service for foreign cars tel: (02) 747 400 or 2422 1635 (Monday to Friday 7am–5pm); nights and weekends 0123.

Car hire
You must be over 21, have a valid licence and have been driving for at least a year. The local firm, **Esocar**, is a little cheaper than most (Husitská 58, Žižkov, tel: 691 22 44).

Documents and insurance
Most national driving licences are valid but an International Driving Licence is advisable. Motorists should bring the vehicle's registration document; green card insurance is also advisable. It is obligatory to carry a first aid kit, a red warning triangle and replacement light bulbs in the car. The vehicle should display a national identification sticker.

Fuel
Petrol (*benzín*) is available as *super* (leaded 96 octane), *special* (leaded 91 octane), *natural* (unleaded 95 octane), *super plus* (unleaded 98 octane), *nafta* (diesel). *Natural* is available in all bigger petrol stations, *super plus* only along motorways.

Traffic regulations
Drive on the right. Seatbelts are obligatory and children under 12 must travel in the back. You are not permitted to drive with any alcohol in the blood. Overtaking trams is forbidden when passengers are boarding and alighting (unless there is a passenger island at the tram stop). Trams have right of way – be alert when crossing tram lines. Speed limits are 110kph on motorways, 90kph on roads and 60kph in built-up areas. The limit for level crossings is 30kph – many have no barriers. Road signs follow European norms. All accidents must be reported to the police (tel: 158 or 02–424141)). Notify accident damage at the Czech Insurance Company, Spálená 14 (tel: 2409 2111).

Men's Suits

UK	36	38	40	42	44	46	48
Rest of Europe	46	48	50	52	54	56	58
US	36	38	40	42	44	46	48

Dress Sizes

UK	8	10	12	14	16	18
France	36	38	40	42	44	46
Italy	38	40	42	44	46	48
Rest of Europe	34	36	38	40	42	44
US	6	8	10	12	14	16

Men's Shirts

UK	14	14.5	15	15.5	16	16.5	17
Rest of Europe	36	37	38	39/40	41	42	43
US	14	14.5	15	15.5	16	16.5	17

Men's Shoes

UK	7	7.5	8.5	9.5	10.5	11
Rest of Europe	41	42	43	44	45	46
US	8	8.5	9.5	10.5	11.5	12

Women's Shoes

UK	4.5	5	5.5	6	6.5	7
Rest of Europe	38	38	39	39	40	41
US	6	6.5	7	7.5	8	8.5

Conversion Table

FROM	TO	MULTIPLY BY
Inches	Centimetres	2.54
Feet	Metres	0.3048
Yards	Metres	0.9144
Miles	Kilometres	1.6090
Acres	Hectares	0.4047
Gallons	Litres	4.5460
Ounces	Grams	28.35
Pounds	Grams	453.6
Pounds	Kilograms	0.4536
Tons	Tonnes	1.0160

To convert back, for example from centimetres to inches, divide by the number in the the third column.

ELECTRICITY

220 volts, 50 cycle AC. Standard continental adaptors are suitable. Visitors with appliances requiring 100/120 volts will need a voltage transformer.

EMBASSIES

Australia Činská 4, Praha 6. Tel: 311 0641 (2431 0070).
Canada Mickiewiczova 6, Hradčany, Praha 6. Tel: 2431 1108.
UK Thunovská 14, Malá Strana. Tel: 2451 0439. The UK embassy acts on behalf of citizens of Ireland and New Zealand.
US Tržiště 15, Malá Strana. Tel: 2451 0848.

EMERGENCY TELEPHONE NUMBERS

Ambulance tel: 333.
Chemist (24-hour) tel: 2421 0229 or 26 81 26 (Na příkopě 7; Nové Město).
Dentist Vladislavova 22, Nove Město. Tel: 2422 7663. Emergency service 7pm–7am and at weekends.
Doctor (emergency service) tel: 155.
Fire brigade tel: 150.
First aid tel: 2422 2521 (Palackého 5 – 7pm–7am and weekends).
Police tel: 158.

The Thomas Cook Worldwide Customer Promise offers free emergency assistance at any Thomas Cook Network location to travellers who have purchased their travel tickets at a Thomas Cook location. In addition any MasterCard cardholder may use any Thomas Cook Network location to report loss or theft of their card and obtain an emergency card replacement. Thomas Cook travellers' cheque refund (24-hour service – report loss or theft within 24 hours). Tel: 44 733 502995 (reverse charges).

HEALTH

Emergency treatment is free. A reciprocal health care agreement exists between Great Britain and the Czech Republic which covers most, but not all, eventualities. US passport holders must pay for treatment. If you need attention apply to the **Fakultní Poliklinika** (2nd Floor) at Karlovo náměstí 28 (open: 8am–4.15pm, tel: 29 93 81, 29 13 53 or 29 79 14). There is also a dental service here and English is spoken. For private treatment you can attend the **Diplomatic Health Centre for Foreigners** at Na Homolce, Roentgenova 2, Praha 5 – Smíchov (tel: 5292 1111). A deposit of at least 1,000kč will be required. It is best to telephone first on 52 60 40 or 5292 2191 out of hours.

Up-to-date health advice can be obtained from your Thomas Cook travel consultant or direct from the Thomas Cook Travel Clinic, 45 Berkeley Street, London W1A 1EB (tel: 071–408 4157). Immunisations against tetanus and polio are recommended. If you intend to stay in forested areas south and west of Prague, north of Brno and west of Plzeň, you are recommended to obtain vaccination against tick-borne encephalitis. This is not necessary for short trips to these areas. Food and water are safe. AIDS is present and likely to increase as prostitution has been a growth industry since the revolution.

INSURANCE

Travel insurance is advisable. Check that the policy covers all medical treatment, loss of documents, repatriation, baggage, money and valuables. Theft may be increasingly difficult to cover. Green Card international insurance is recommended for your car.

LANGUAGE

Czech sounds and looks daunting, but English is fairly widespread among the post-war generations; most older people speak some German.

Pronunciation

The stress is always on the first syllable of a word.

Vowel sounds

a like the English '**u**' in '**up**'.
á as in 'r**a**ther'.
e as in 'g**e**t'.
é similar to '**ai**' in 'h**ai**r'.
ě '**ye**' as in '**yes**'
i as in 's**i**t'.
í or **y** as in 'm**ea**t'.
o as in 'h**o**t'.
ó as in 'cl**aw**'.
u as in 'b**oo**k'.
ů or ú as in 'c**oo**l'.

Dipthongs

au as in '**now**'.
ou as in '**oh**'.

Consonants

Consonants are the same as in English with the following exceptions:
c as in 'oa **ts**'.
č as in ' **ch** ess'.
j as in '**y** ou'.
ch as in Scottish 'lo **ch**'.
r like English but rolled.
ř rolled and combined with **zh** sound in 'plea **s** ure'.
š as in '**sh** e'.
ž like the **zh** sound in 'plea **s** ure"'
d, **t** and **n**, if followed by **i** or **í**, become **dyi**, **tyi** and **nyi**.

Numbers

1	jeden	**6**	šest
2	dva	**7**	sedm
3	tři	**8**	osm
4	čtyří	**9**	devět
5	pět	**10**	deset

Basic phrases

yes	ano
no	ne
please/ you're welcome	prosím
thank you	děkuji
bon appetit	dobrou chut
hello	ahoj
goodbye	na shledanou
good morning	dobré ráno
good day	dobrý den
good evening	dobrý večer
good night	dobrou noc
small	malý
large	velký
quickly	rychle
slowly	pomalu
cold	studený
hot	horký
left	nalevo
right	napravo
straight ahead	přímo
where?	kde?
when?	kdy?
why?	proč?
open	otevřeno
closed	zavřeno
how much?	kolik?
expensive	drahý
cheap	levný
near	blízko
far	daleko
day	den
week	týden
month	měsíc
year	rok

Days of the week

Monday	pondělí
Tuesday	úterý
Wednesday	středa
Thursday	čtvrtek
Friday	pátek
Saturday	sobota
Sunday	neděle.

LOST PROPERTY

Apply to Bolzanova 5, Nové Město (tel: 2422 6133).

MAPS

Tourist Information Offices (see page 189) and Chequepoint Exchange kiosks issue free street plans. The most comprehensive and up-to-date maps are produced locally by **Kartografie Praha** (1:20,000). They include public transport routes.

MEDIA

Prague boasts two excellent English language papers: *The Prague Post* (weekly) and *Prognosis* (fortnightly). They have listings, restaurant reviews and other indispensable information. Together with English, European and some American papers, they can be bought at news-stands around the centre. Radio 1 (91.9 MHz) has a news bulletin in English at 3.30pm on weekdays. There are four TV channels: ČT1, ČT2, ČT3 and Premiéra TV. ČT3 is devoted to western services including CNN and BBC.

A facelift for the Czech crown

MONEY MATTERS

The Czech unit of currency is the *Koruna Ceska* (Czech crown, abbreviated to kč). It is divided into 100 *haléra* (hellers) in denominations of 10, 20 and 50h. There are 1, 2, 5, 10 and 50kč coins and 20, 50, 100, 200, 500, and 1,000kč notes. New notes for 20, 50 and 5,000kč are planned. If receipts are retained, surplus crowns can be converted back to hard currency at the end of your stay. Import and export of over 100kč is illegal.

Changing money

Exchange kiosks charge high commission and many indulge in suspect practices. The banks tend to have long queues in the high season, but most give better value. The Živnostenská Banka (Na příkopě 20) is unbureaucratic, light on charges and open from 8am to 6pm Monday to Friday and 8am to noon on Saturday.

Travellers' cheques and Eurocheques are widely accepted at banks and exchange counters if issued by a well-known bank. Credit cards can be used to obtain cash advances from the larger banks and exchange kiosks; they are also accepted in some luxury shops and expensive restaurants.

Usual banking hours are 8am–2pm, Monday to Friday. Some banks in the city centre have longer hours and close over lunch. The airport exchange desk is open 24 hours.

Thomas Cook MasterCard

travellers' cheques free you from the hazards of carrying large amounts of cash and in the event of loss or theft can quickly be refunded (see **Emergency Telephone Numbers**). US dollar cheques are recommended, although cheques denominated in major European currencies are accepted. Many hotels, shops and restaurants in tourist areas accept travellers' cheques in lieu of cash.

Thomas Cook Czechoslovakia a.s. on Wenceslas Square (Václavské namĕstí 47 – tel: 2422 9537), can provide emergency assistance in the case of loss or theft of Thomas Cook MasterCard travellers' cheques and will change currency and cash travellers' cheques (free of commission in the case of Thomas Cook MasterCard travellers cheques). The branch at Motokov Building, Na Strži 63, Prague (tel: 6114 2912) is also able to provide emergency assistance.

OPENING HOURS

Department stores and food shops – 8am–6pm, Saturday to midday. (Some shops, such as supermarkets, may have longer hours).

Galleries and museums – Tuesday to Sunday 10am–5pm (in some cases 6pm or 7pm). All the sights comprising the Jewish Museum in Josefov are closed on Saturdays.

POLICE

Look for *Policie*. The Central Police Station is at Konviktská 14 (tel: 2413 1111. Emergencies 158). For visa extensions and residence permits go to Olšanská 2, 3rd District (open: Monday to Friday 8am–noon. 12.30–3.30pm, but closed Wednesday afternoons.

POST OFFICES

The Central Post Office and Poste Restante are at Jindřišská 14, Praha 1 (tel: 2422 8856 or 2422 8588). Stamps and phone cards can be purchased both here and from newsagents/ tobacconists.

PUBLIC HOLIDAYS

1 January – New Year's Day;
Easter Monday – (movable);
1 May – Labour Day;
8 May – VE Day;
5 July – Day of Slovan Missionaries Cyril and Methodius;
6 July – Jan Hus State Holiday;
28 October – Independence Day;
25–6 December – Christmas.

PRAGUE METRO

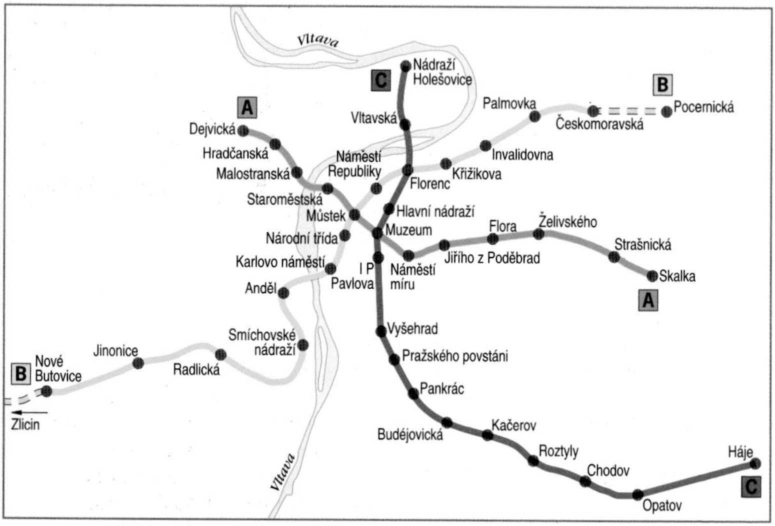

PUBLIC TRANSPORT

Public transport (see pages 22–3) is cheap and efficient. Separate tickets are required for each part of a journey involving changes, except on the metro. The Five-Day Tourist Ticket is valid for all public transport. The three-day Prague Card combines a season ticket with free entry to major sights and some museums. (Apply at Čedok, Na příkopě 18, Pragotour, U Obecního domu 2, American Express, Václavské náměstí 56.) Public transport operates generally between 5am and midnight. Night trams, covering most of Central Prague, all stop at Lazarská, close to Wenceslas Square. They leave every 40 minutes between 11.30pm and 4.30am.

Central Bus Station: Křižíkova 4 (Metro to Florenc). Tel: 2421 1060 (for timetable information).

It is best to avoid taxi ranks in the town centre and call a reliable taxi

company. Taxis can be summoned on: 35 03 20, 34 24 10, 2491 1559 or 2491 2344.

Outside Prague

The rail network connects Prague with many other towns within the country, as well as other major German and East European cities. For further details and timetables, consult the *Thomas Cook European Timetable*, published monthly and obtainable from branches of Thomas Cook.

RELIGIOUS WORSHIP

Mass and confession in English are held at St Joseph's Church (Josefská, Malá Strana) every Sunday (confession from 10am, mass at 10.30am). Otherwise masses are in Czech, as are the Protestant services at St Nicholas on Staroměstské náměstí.

SENIOR CITIZENS

Over-70s travel free on public transport.

STUDENT AND YOUTH TRAVEL

The **CKM-SSM** (Youth Travel Agency) is at Žitná 10 Praha 2, where there is also a **Juniorhotel** for under 30s (tel: 2491 5767).International Student Identity Cards are issued at the CKM office on Jindřišská 28 (tel: 2423 0218). See **Hotels and Accommodation**, pages 174–7.

TELEPHONES

Modern phone booths that accept telephone cards are to be found all over Prague. The yellow and black phones that take 1kč are only for local calls; long-distance calls are made from the grey ones (1, 2, and 5kč). Most have instructions in English. Call the English speaking operator on 0135. The dialling code from Prague for the **UK** is 0044, for **Eire** 00353, for **Australia** 0061, for **New Zealand** 0064, for **USA** and **Canada** 001.

TIME

The Czech Republic follows Central European Time which is Greenwich Mean Time + 1 hour, or US Eastern Standard Time + 7 hours. Between March and late September clocks are advanced 1 hour (GMT + 2 hours). Prague is four and a half to 10 hours ahead of Canada, seven to nine hours later than Australia and 11 hours later than New Zealand.

TOILETS

Toilets (*záchod*) are few and far between, but metro stations quite often have one. The sign is WC or *Muži* (men) and *Ženy* (women). In women's loos, and usually in men's, you will have to pay a few kčs.

TOURIST INFORMATION

The Prague Information Service (Pražská informační služba – PIS) has its main office at Betlemské náměstí 2, Nové Město, Praha 1. Open: Monday to Friday 8am–7pm, Saturday and Sunday 8am–3.30pm, (later in high season). There is another convenient office at Staroměstské náměstí 22, and on Hlavní nádraží (Central Railway Station) and yet others scattered strategically around the city. For general information ring 54 44 44.

ACKNOWLEDGEMENTS
The Automobile Association wishes to thank the following organisations, libraries and photographers for their assistance in the preparation of this book.
PETR BALAJKA 156, 157; INTERNATIONAL PHOTOBANK cover; PRAGUE NATIONAL GALLERY 76/7, 77, 91; REX FEATURES LTD 12 (T Haley), 13; ZEFA PICTURE LIBRARY (UK) LTD inset.
The remaining photographs are held in the AA Photo Library and were taken by Jon Wyand with the exception of pages 1, 6, 24, 26, 27, 36, 42, 48, 50, 51, 56, 57, 58, 59, 61b, 63, 75, 79, 89, 90, 92a, 92b, 97, 98, 102, 105, 108, 109, 112, 131, 151, 160, 165 and 188 which were taken by Antony Souter and page 125 which was taken by Clive Sawyer.
The author would like to thank Pavel and Alice Stecha and family, Radovan Bocek, Ztenka Bocskova and Vera Behal for much kindness and practical assistance.

The Automobile Association would also like to thank Mr Jiri Krejca and Mr Stewart Marshall, Thomas Cook, Prague, and Martin Velik, UAMK (Prague) for his invaluable assistance.

Series adviser: Melissa Shales

Copy editors: Audrey and Terry Horne